ADVANCES IN
WATER TREATMENT
AND
ENVIRONMENTAL
MANAGEMENT

Proceedings of the
1st International Conference
(Lyon, France 27-29 June 1990)

ACKNOWLEDGEMENTS

The valuable assistance of the Technical Advisory Committee and panel of referees is gratefully acknowledged.

TECHNICAL ADVISORY COMMITTEE

Mr G A Thomas (Chairman)	Thames Water Authority
Mr R Hyde	Water Research Centre
Mr D Latham	Anglian Water Authority
Professor D G Stevenson	PUT Projects Ltd
Mr R F Stoner	Mott McDonald Group Ltd
Mr D Wilson	Yorkshire Water Authority
Mrs L Grove	BHR Group Ltd

OVERSEAS CORRESPONDING MEMBERS

Mr V van den Bergen	Dutch Ministry Water Department	Netherlands
Mr T Burke	Water Research Centre Inc	USA
Mr J L Trancart	Department for Water Treatment	France

ADVANCES IN WATER TREATMENT AND ENVIRONMENTAL MANAGEMENT

EDITORS

George Thomas
Consultant

Roger King
BHR Group Ltd, Cranfield

CRC Press
Taylor & Francis Group
Boca Raton London New York

CRC Press is an imprint of the
Taylor & Francis Group, an **informa** business

Organised by BHR Group Limited and co-sponsored by the
Commission of the European Communities

CRC Press
Taylor & Francis Group
6000 Broken Sound Parkway NW, Suite 300
Boca Raton, FL 33487-2742

First issued in paperback 2019

© 1991 by Taylor & Francis Group, LLC
CRC Press is an imprint of Taylor & Francis Group, an Informa business

No claim to original U.S. Government works

ISBN-13: 978-1-85166-632-4 (hbk)
ISBN-13: 978-0-367-86596-2 (pbk)

British Library Cataloguing in Publication Data

Advances in water treatment and environmental management: proceedings of the 1st International Conference.
1. Water. Quality
333.91

Visit the Taylor & Francis Web site at
http://www.taylorandfrancis.com

and the CRC Press Web site at
http://www.crcpress.com

CONTENTS

PART VI—RIVERS AND RIVER MANAGEMENT

PART VII—MANAGEMENT OF ESTUARIES AND BEACHES

ADDRESS

The theme of the conference – progress in the field of managing the environment – is particularly interesting to the people of Lyon. Courly may justifiably lay claim to the privilege of being an economic metropolis within a remarkable environment.

Two subjects exemplify (especially the supply of drinking water) the Lyonese sensitivity to environmental problems.

1 SUPPLY OF DRINKING WATER

The initial security is assured through a perennial resource, both quantitatively and qualitatively. Quantitatively, the drinking water supply of the urban community of Lyons is for the most part guaranteed by the great alluvial water table of the Rhône; production may further mobilise re-supply reservoirs. Although equipped since 1976 with statutory protection perimeters (400 hectares of which are in the immediate area), this catchment zone remains a tributary of the river. The situation requires very special viligance over the entire upper waters of the Rhône to ensure that the risk of accidental pollution (industry and roads) does not jeopardise the functioning of this resource.

Then, the new risks linked to economic and industrial development must be overcome. How do these affect Lyon? First, there is the permanent requirement for an ever greater security. The increasing complexity of our city makes it all the more vulnerable. We know the risks involved for a large conurbation of a long-term break in the water supply.

Searching for an answer to this problem gave rise to a universal re-think concerning the security of the supply, to an original initiative and finally to an emergency water treatment plant. This plant is solely intended to operate in the event of accidental pollution of the Rhône, constituting complete alerting mechanism (comprising a list of risks, of equipment for the real time detection of pollution, etc.).

Today, this concern to improve security is taking priority over policies designed to increase the capacity of the installations: 90% of our investment is investment in security and redundancy.

2 On the other hand, solving technical problems relating to the environment has always been accompanied by the will to develop technologies taken on in a close partnership with companies which are working with Courly.

3 Similarly, in the area of sanitation, which, contrary to that of drinking water, has not progressed in the course of recent decades, Courly has developed new technologies and new operating equipment. This includes:

- the construction of two fixed culture purification plants: one at Meyzieu with a capacity of 35 000 Eq/inhab. and one at Fontaines-Sur-Saone with a capacity of 40 000 Eq/inhab. approximately. Apart from their advanced technology and their fundamental water treatment role, these two plants have been designed with two essential objectives: protecting the environment (odours, noise, appearance) and major facilities made available to the operator enabling him to control and regulate the functioning of the plant and its purification performances.
- the development of an expert system to assist decision-making at the Meyzieu purification plant (ORAGE). This equipment offers the plant operator guidance when making his major decisions (flow-rate to be treated, number of works in operation, preventive maintenance, sludge treatment), no longer according to currently available resources and common practice, but on the basis of rational data and real measurements edited and stored by the system.
- setting up the GESICA system (Management, Simulation and design of Sanitation). This system, based on the computerised urban cartography on which the sanitation network is located, enables the design office and the network operator to obtain assistance in the planning of investment (network calibration, hydraulic load) and rehabilitations (condition of structures), to rationalise the sewage purification campaigns and to act rapidly in the event of accidental pollution.
- installing equipment for the remote management of networks and purification and detection plants (TERESTA). This equipment centralises, in real time, the data from 60 plants (purification and detection), thereby making it possible to take very rapid action with respect to operating faults, to carry out preventive maintenance on the installations, to manage operating costs and to bring operating personnel a great deal of comfort and flexibility in their work.

These innovations are put into practice in a permanent spirit of partnership between the municipality of Courly, private enterprise, the university and the 'grandes écoles', who are pooling their know-how, their experience and their respective resources.

These efforts clearly contribute to the accomplishment of the tasks of the Sanitation Services, which are the collection, transportation and treatment of waste water. In addition, the Sanitation Service is increasingly sensitive to the necessity of dealing with rainwater and the pollution that it entrains.

C Mansat
Sanitation Service
Lyon. June 1990

FOREWORD

The quality of the environment, and particularly of water, is becoming an increasingly important issue around the world and especially so in Europe, where the quality of each nation's water is increasingly judged against standards established in EC Directives, and where the water sector is undergoing change due to factors such as privatisation.

This volume contains a discussion of developments in policy, current practice and emerging technologies in the water sector and related industries and the environment. The papers in this collection are grouped into sections covering: evolution of policy, groundwater quality, plant design and construction, plant development, control and measurement techniques, and water treatment technologies and applications. The interaction of water with the wider environment is discussed in sections devoted to rivers and river management and to the management of estuaries and beaches.

Participants in the discussion on policy matters include a spokesman for the National Rivers Authority addressing the role of this new regulatory agency for the UK's water, and a spokeswoman from Compagnie Générale des Eaux on the challenges of implementing the EC's drinking water directive.

Papers devoted to design and construction of plant and plant development discuss how the water sector is being increasingly drawn towards the process industries both in terms of approach and of technology. Specific case studies of installations at Lyon are presented by local representatives from the Compagnie Générale des Eaux and the Greater Lyon Council.

Reports on new work on specific water treatment technologies include the use of zeolites as a possible alternative to granular activated carbon for removing chlorinated organics from drinking water (performed in the School of Chemical Engineering at the University of Bath, UK), a computer model used to improve ion exchange denitrification of drinking water at high sulphate levels (at the Cranfield Institute of Technology, UK) and modelling of the efficiency of hydrocyclones in treating oily water (Centre d'Etudes et de Recherche at Grenoble, France).

With regard to the interaction of water with the wider environment, keynote speaker (and prize-winning environmental author) Jeremy Pursglove stresses the positive contribution engineers can make when given a wider 'greener' brief. Pursglove explains that the straightening out of rivers and stripping of trees from their banks resulted from engineers being given too narrow a brief which did not include the potential for making environmental improvements. Environmental considerations can even pay their way, says Pursglove, citing the example of 'science parks' which are well landscaped and command a higher price. Pursglove denies that he was fighting an uphill battle when trying to introduce 'green' concepts into engineering schemes—'though it is still a battle'—but he stresses that such ideas must be brought in at the initial stage to avoid costly changes later.

A summary of trends affecting the water sector is given by François Fiessinger (formerly of Lyonnaise des Eaux and now Vice President of Zenon Environmental Inc. in Canada) in a keynote speech. Fiessinger identifies the following as the strongest trends influencing the water sector: the drive for higher quality water, particularly in the USA with the Safe Drinking Water Act; progress in technologies such as electronics, biotechnology, polymer science and membrane separation with the consequent demand for 'high-tech' to be applied to water; and the development of privatisation, particularly in Europe, resulting in increasing competition between water organisations. This will encourage former public services to look for new markets and to develop a policy of diversification, with the most common area being water treatment. This in turn will result in 'large' organisations.

According to Fiessinger, the conjunction of several of these forces could create 'windows of opportunity' for specific technologies in the 1990s. He sees the biggest potential in the use of fast-improving membrane technologies for treating drinking water and the application of biological methods (such as anaerobic treatment and biofilters) to the treatment of wastewater effluents. Fiessinger is an enthusiastic proponent of membrane technologies and foresees the combination of membrane separation with biotreatment into fully integrated membrane bioreactors bringing '. . . compactness, higher yields, reliability, ease of operation, low sludge production and altogether a much better quality water'.

The views and information given here will help all readers to be better placed to realise the opportunities which new organisations and technologies might bring and to contribute to improving the quality of Europe's water and associated environment.

John O'Hara
Environment & Industry Digest
December 1990

PART I

Policy matters

Chapter 1

ENVIRONMENTAL STEWARDSHIP: THE ROLE OF THE NRA

P A Chave (National Rivers Authority, London, UK)

INTRODUCTION

The National Rivers Authority came into existence on the 1st September 1989. It was set up under the Water Act, as an independent watch dog, separated from its former Water Authority parentage, as a result of pressure from a variety of sources to recognise the increasing need to separate the regulator from the regulated, and from the perceived need for a body to care for the environment - in the NRA's case the aquatic environment. The title of this lecture reflects the view that the environment does not look after itself - it has to be watched over by a caring and committed organisation. Within the United Kingdom over the past few decades a variety of public bodies have been entrusted with the care of the natural freshwater network of rivers. Under the Rivers Prevention of Pollution Act 1951, Rivers Boards were set up in England and Wales, to be followed by River Authorities on enactment of the 1963 Water Resources Act, and latterly Water Authorities, operating through a multi-functional water-cycle approach, have had this responsibility following the enactment of the Water Act 1973. In Scotland, different legislation led to the establishment of River Purification Boards, and in Northern Ireland, the Department of the Environment (NI) assumes responsibility for such matters.

The NRA has been given a wide range of responsibilities: constantly in the public eye are those related to the prevention of pollution and the maintenance of water quality in controlled waters, but its other duties are equally important. The management and provision of water resources is a matter of considerable concern and public interest at the present time, with apparent changes in climatic conditions giving rise to the potential for regular shortages of water in certain parts of the country. The controlled allocation of abstraction of water from rivers and aquifers is an important matter for the NRA which has duties to ensure the availability of both water

© 1991 Elsevier Science Publishers Ltd, England
Water Treatment – Proceedings of the 1st International Conference, pp. 3–16

for public supply, and water for those other abstractors requiring it. The maintenance of a balance of abstraction versus the minimum required flow in our rivers has an important bearing upon the environmental quality of those waters. Whereas at the present time we have concerns over the availability of adequate water for our needs, the NRA also has responsibility at the other end of the scale relating to water quantity, that is flood prevention. In recent months, we have experienced excessively wet conditions in the UK, which has thrown an emphasis onto our work in this arena - particularly in relation to our role in maintaining and operating flood defence systems, for the control of inland flooding from some of our major water courses and those relating to sea defences. Moving on again to further responsibilities put upon the NRA by the Water Act, it has duties to promote conservation and recreation on waters and associated lands to which it has rights. It has a duty to develop salmon, trout, freshwater and eel fisheries and in some areas of the country the NRA is a navigation authority.

This brief introduction has defined the breadth of the NRA's responsibilities, all connected with the maintenance of the natural aquatic environment.

GEOGRAPHICAL AREA OF RESPONSIBILITY

In view of the subject matter of this conference, I propose to concentrate on the role of the NRA in controlling pollution and enhancing the quality of our natural waters. This cannot be separated entirely from the business of controlling water resources and the reduction of excessive abstraction.

The 1989 Water Act conferred specific responsibilities on the NRA for water quality in all controlled waters, and for the achievement of water quality objectives which are to be determined by the Secretary of State for the Environment and for which the NRA will be required to develop the related water quality standards.

Definitions of the extent to which its duties and powers apply, in geographical terms, are laid down in the Water Act and such other legislation as was carried through to this Act - for example its flood prevention activities on inland water relates to a definition of "main river"; pollution control and environmental monitoring activities are restricted to "controlled waters". An important aspect of the developing legislation is the ever increasing range of waters which have been brought under control, from freshwater rivers in the 1951 Act, to the definition of controlled waters now set out in the Water Act 1989.

Controlled waters are defined in Section 103 of the Act and effectively cover all surface waters including rivers and, by a subsequent Regulation issued by the Secretary of State, public supply reservoirs; estuaries; and the sea, to a distance of 3 miles from the shore; and virtually all underground waters. It follows from this that treated water supplies are not within the NRA's responsibility. The quality of _treated_ water supplied to consumers is the responsibility of the Water companies, and its quality is controlled by separate Regulations - the Water Supply (Water Quality) Regulations 1989 - which incorporate the requirements of the EC Drinking Water Directive together with sampling regimes which are strictly defined.

So, for all practical purposes, any naturally occurring water within England and Wales will come within the jurisdiction and stewardship of the NRA.

CLASSIFICATION SCHEMES

In the past the assessment of water quality and public reports upon this have been restricted to a detailed examination of river quality in England and Wales using a classification scheme devised by the former National Water Council in 1978. This was extended in 1980 to include estuaries, but the classification used in this case was of an elementary nature. The classification scheme defined stretches of rivers as falling into different categories and these were based on some specific parameters, but also took into account the current potential uses of the waters. This system had no statutory base and did not relate directly to EC Directives. The definition of controlled waters extends to a much wider range of waters the need to devise and use a classification and assessment scheme. Under the Water Act any such scheme will have a statutory status through Section 105 of the Act. At the present time the NRA and the Department of the Environment (DoE) are working on the preparation of new classification schemes with the objective of providing an absolute comparison of water quality which is appropriate to the nature of the water between different water courses, different regions, and different years, covering all types of waters.

WATER QUALITY OBJECTIVES

Section 104 of the Water Act also allows the Secretary of State to set water quality objectives for controlled waters, and to issue a date by which waters must satisfy these objectives. The use of water quality objectives has, for many years, been the central basis through which the UK system of water pollution control has operated. Nevertheless, until now any water quality objectives which have been used have been non-statutory in character. Once the Secretary of State has set such water quality objectives it will be the

NRA's duty to ensure that as far as is practicable the water quality objectives are achieved. The NRA is considering the criteria which will be necessary in order to establish water quality objectives and is currently creating categories for the various uses to which water may be put, in order to be able to derive specific water quality standards appropriate to each use-related objective.

WATER QUALITY SURVEYS

River water quality has been reviewed on a quinquennial basis since 1958 under a scheme operated by the Department of the Environment. The next such review is due in 1990. The NRA has now embarked on a two-part review - the first part a survey designed to replicate as closely as possible those previous surveys. It has been recognised however, that earlier surveys suffered from the differences in approach adopted by the former River Authorities and Regional Water Authorities, and, furthermore, that little attention was given to biological criteria. For these reasons the NRA is conducting an overlapping survey based on a standard set of procedures and incorporating a biological survey making use of a river invertebrate predicting and classification model (RIVPACS), which has been developed by the Institute of Freshwater Ecology, through funding from the DoE and the Natural Environment Research Council. The use of such a system will enable a biological override to be attached to the results of chemical analysis to give a more reliable assessment of water quality which relates to the topographical and other features of the assessment point.

The results of the 1985 survey of river quality which were published by the DoE showed that 90% of rivers included in the classification were generally of satisfactory quality - in classes 1 and 2 of the old classification system - but the report identified marked regional variations. The largest concentrations of polluted rivers were located in the densely populated and industrialised areas of England and Wales. Major problems relating to the discharge of untreated sewage were identified in the Mersey and Humber rivers. River quality problems in England and Wales were mainly connected with the quality of effluents from sewage treatment works, and pollution from intensive agriculture and forestry. A significant conclusion from the last survey was that the reductions in the lengths of seriously polluted rivers and canals which had been achieved in the 1970s had not been maintained in the early 1980s. Whilst in some regions improvements were continuing, there had been a small net deterioration in overall quality between 1980 and 1985. The results of the 1990 survey, which will be carried out on comparable data and by similar assessment techniques, will be extremely interesting. This report is due to be published towards the end of 1991.

DISCHARGE CONSENTS

Arising from the duties of the NRA to maintain and improve water quality is a requirement to control and remedy pollution. The Water Act in Sections 107 and 108 sets out some of the major powers available to the NRA to achieve this objective.

The main thrust of the NRA's ability to act arises from the fact that, except for consented discharges, it is an offence to cause or permit poisonous, noxious or pollution matter or solid waste matter to enter controlled waters, and the NRA is able to prosecute those who commit such offences. It is clear that the NRA intends to use these powers to the utmost, an example being the vigorous prosecution of Shell U.K. following an oil pipeline fracture under the River Mersey which led to the imposition of a £1M fine.

Effluents from industrial firms and sewage treatment works in liquid form are disposed of to water courses, and under Section 108 of the Act, such discharges are permitted provided a consent has been obtained from the NRA. The consenting system has operated for many years under previous legislation but it became apparent, soon after the NRA was formed, that it had inherited a wide variety of such consents. Very large numbers of consents had been issued by the former organisations, River Boards, River Authorities and Water Authorities and many been changed according to various criteria over the years. For example, in the late 1970s and early 1980s many existing consent conditions were "rationalised" in order to reflect the current performance of sewage works. Furthermore, previous legislation had not covered such a wide variety of waters as are now designated "controlled waters" with the result that upon the enactment of the 1989 Water Act, many discharges which were previously outside of statutory control, were rapidly brought into such control and had to be given "deemed" status pending the determination of conditions. In other words, such discharges were declared legal but with no conditions attached. These discharges included many coastal outfalls, which were exempted from control until 1987 and a large number of previously illegal discharges, including storm sewage overflows, which required consents upon the enactment of the Water Act.

The proliferation of such a variety of consents is clearly undesirable and taking into the fact that some £20 bn of expenditure has been committed to the improvement of existing discharges over the next few years by the water industry, some consistency of approach is highly desirable. Consequently the Secretary of State asked the NRA to set up a policy group to examine these issues and this will be reporting in the spring of this year. The group is considering such matters as how the regulator and discharger should assess whether a particular discharge meets the terms of its consent; the extent to which consent requirements for different types of discharge should be put on a common basis and how to justify any differences; how best to ensure that discharges are properly classified as compliant or non-compliant; sampling frequencies; and the role of consent conditions in relation to polluting load.

Information relating to consents is held on a public register as are the results of monitoring both discharges and the aquatic environment in general.

POLLUTION INCIDENTS

Whilst the NRA's main thrust of pollution control is likely to be through the use of the discharge consent system, polluting events will always occur. In order to reduce the effects of these upon the aquatic environment, and the disruption to legitimate users of water that this entails, the NRA is also empowered to take action to both mitigate the consequences and take such preventative action as is available to it.

The total number of water pollution incidents in England and Wales has demonstrated a rising trend since 1980 (Figure 1); the number of such incidents in 1987 was 80% higher than the equivalent figure in 1980. 37% of these incidents were attributable to industry, 19% to farm pollution and 20% to sewage and sewerage problems (Figure 2). Of the industrial incidents, oil pollution accounted for 61% of the total number. Increased public interest, and increased monitoring can confidently be expected to bring about a further increase in these figures in the current year and beyond. In order to bring about a reduction in such pollution events, certain actions are available now under the Water Act to enable the NRA to take preventative action. The

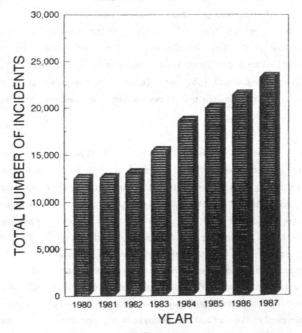

WATER POLLUTION INCIDENTS
ENGLAND AND WALES

Figure 1

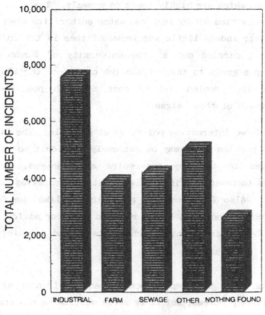

WATER POLLUTION INCIDENTS
REPORTED IN ENGLAND AND WALES FOR 1987

Figure 2

Secretary of State may designate water protection zones for sensitive areas under Section 111 of the Act with a consequent prohibition of certain activities within those areas. The NRA is also entitled, under Section 115, to undertake works to deal with an incident as it occurs and indeed to undertake works to prevent a perceived problem occurring, and to charge the actual or potential polluter.

An interesting example of the use of water protection zones are so-called nitrate sensitive areas. There is currently considerable concern over nitrate levels in drinking water, and a general perception that such levels are rising in groundwater and surface water. In order to offer a general remedy to reduce nitrate levels, the Water Act permits the designation of nitrate sensitive areas in which agricultural practices can be modified. A pilot scheme has been set up by the Ministry of Agriculture, Fisheries and Food in conjunction with the NRA to run for 5 years. Ten sites have been identified, and we will look for major changes in agricultural practice in these areas to assess whether such changes do represent a suitable means of reducing nitrate levels.

This may be of relevance in due course to the problem which is currently being experienced throughout the UK and elsewhere in the world concerning development of blue-green algae. Within weeks of being set up, the NRA was confronted with a so-called blue-green algal bloom phenomenon. The species concerned included <u>microcystis aeruginosa</u> which is known to be capable of releasing substances which are highly toxic to mammals. Toxicity testing had not previously been carried out by regional water authorities when such blooms occurred in the past; indeed little was known of them in the United Kingdom except through work carried out at the University of Dundee. The NRA subsequently set up a group to investigate the causes of the 1989 bloom, to review the world-wide problem and to come up with possible means of controlling the growth of these algae.

This subject raised two interesting points about our role. The first is that we may be facing a problem which may be extremely difficult to solve without fundamental changes to the way reservoirs are managed, or involving significant changes to sewage and industrial treatment processes, and possibly farming practices. Also it raised the point that unlike the former Water Authorities, the NRA has no legal basis on which to offer advice relating to health matters. The NRA is an environmental agency and is solely concerned with environmental factors.

This conference is about treatment processes and environmental management rather than farm waste problems, but it is worth mentioning two aspects of our current problems from the agricultural industry in the UK. It is estimated that there are some 200 million tons of animal wastes and the effluent from 36 million tons of silage to be dealt with each year. There are over 200 different chemicals in use or stored on farms and all too often, these wastes and chemicals find their way into water courses. Recent catchment surveys in several regions, have shown that pollution incidents arising from agricultural is far more widespread than indicated by the reported incidents - which rose dramatically from 1979 to 1988 and which relate largely to the more visible organic contamination of natural waters. In some surveys in intensive dairy catchments, up to 25% of farms were found to be polluting at the time of inspection and a further 25% were at risk of doing so. Such substances cannot always be dealt with on a daily basis and often the entire slurry production must be contained for all of the winter period requiring stores of $2000M^3$ or more. The potential environmental impact of such stores failing is thus very large and the DoE intends to reduce Regulations this year under Section 110 of the Water Act to control the storage of such material in the farming situation. This is a welcome move and the DoE are considering preparing similar Regulations for industrial sites.

The other aspect of farming which should be mentioned relates to the general use of pesticides. Although adequate controls exist under the Water Act to control the discharge of point source emissions of pesticides, it is much more difficult to control diffuse entries. The variety of pesticides used, the necessity to develop methods applicable at a very low concentrations and the cost of analysis limits the number of water samples analyzed. An attempt is now being made to address this problem in two ways. The proposal to incorporate biological methods of assessing water quality using RIVPACS will identify those reaches of river of apparently good quality but which have a reduced biological quality and this may help to locate problems arising from the discharge of chemicals, including pesticides. We are increasingly using analyses of pesticide levels in sediment and biological material such as fish tissue to obtain a long-term indication of water quality. Work to assess the industrial production and discharge of pesticides under the surveys of Red List substances currently under way should provide further evidence.

A possible approach to the control of diffuse sources of pollution of this kind may be to limit the use of difficult substances in catchments by the use of water protection zones.

EC DIRECTIVES

So far this paper has been concerned the position relating to internal legislation in England and Wales. As members of the European Community, the UK is also subject to a large number of regulations concerning the aquatic environment set out in EC Directives. The NRA has generally been given the task of operating such Directives. Directives encompass the complete range of waters for which the NRA has environmental stewardship, from the Bathing Water Directive currently applying only to coastal sites in the UK, the Freshwater Fish Directive, applicable to rivers and certain large lakes, to the Groundwater Directive, which covers underground controlled waters.

In general Directives of this nature have been brought into use within the UK through the Departmental Circular. The European Commission has recently indicated that, in their view, such a route for implementation is unacceptable, in that all Directives are binding upon member states, and must be incorporated within their national legislation. In view of this, the Department of the Environment is currently issuing Regulations through the medium of the Water Act to enact Directives.

This action in itself is highlighting problems relating to the implementation, in a formal way of such Directives. For example the NRA is currently in discussion with the Department over implementation of the Directive concerning the abstraction of surface water for the production of drinking water. The majority of surface water supplies in the UK, arise from reservoirs. The classification scheme laid down by the Directive relates the degree of treatment to the quality of water abstracted. In a typical lake where stratification occurs as a natural process, the water at the surface will be different to that in the bottom layers. It would not be possible to guarantee that all parts of a particular lake, at all times, met the classification criteria and indeed in the design of reservoir systems provision is normally made to abstract water from the most suitable level or site. The problem for

the NRA is how to define the point of application in such a way as to achieve
a practical means of monitoring and operating the Directive. It would be
difficult to be required to guarantee that all water in a particular lake was
of precisely the same quality all of the time.

Whereas some of the Directives are straight forward and merely give rise to
a workload for the NRA to accomplish, other Directives are less easy to deal
with. The Dangerous Substances Directive, for example is a Directive with
constantly changing analytical requirements. The transfer of substances which
are toxic, persistent, and non-biodegradable into what is known as List 1 has
been accelerated in recent years, and the need to identify sources of such
substances and provide the analytical capability to detect them at very low
concentrations in natural waters is a challenge to the NRA's new laboratory
network.

COASTAL WATERS

So far the paper has concerned itself with inland problems, but the NRA has
responsibilities for coastal waters also. Two EC Directives are of topical
interest in this context - the Bathing Water Directive, and the proposed
Directive relating to the Treatment of Municipal Waste Water. Both apply to
coastal and inland waters, but within the United Kingdom, no freshwater
bathing sites have yet been designated. Much of the interest in the
application of the Bathing Water Directive to coastal waters centres on the
levels of bacteria and viruses and the implications that these have for
emersion sports. Most inland waters receive inputs of sewage effluent which,
although designed to reduce the quantities of organic material discharged, and
thereby affording reductions in bacteria and viruses, nevertheless in no way
produce a sterile effluent. Receiving waters therefore contain considerably
higher levels of both bacteria and viruses compared with coastal waters, and
these can be substantially reduced only by introducing disinfection
techniques. The NRA is currently examining the merits and problems of such
techniques as applied to marine outfalls and will use this information to
explore their possible application to freshwaters should this prove necessary,
and, as guardian of the water environment, will ensure that any end products
are environmentally acceptable.

The second Directive, which will be of particular interest to this conference,
is the Municipal Waste Water Directive which will be expected to move from the
proposed stage to a become a confirmed Directive later this year. In this
Directive it will be necessary to apply treatment to those sea outfalls which
are currently discharging untreated sewage to sea. The NRA's role in
dealing with this Directive will be to issue consents to discharge which

specify the quality of the effluent in such a way as to ensure that treatment works are installed. One of the difficulties associated with the Directive is the requirement to ascertain whether primary, secondary or tertiary treatment is required, and this will depend on the classification of the water as being sensitive or less sensitive. This is clearly a role for the NRA to perform in due course. Whilst in general the NRA welcomes this Directive it has expressed the view that properly designed long sea outfalls are an acceptable option for discharges into the sea. Furthermore the imposition of fixed standards for effluent quality through the Directive is possibly counter to the use of water quality objectives as an approach.

In addition to work relating to EC Directives, the NRA has a number of international obligations through its association with the North Sea Programme, and various international conventions.

RESOURCES AND RESEARCH

It will be clear from the variety of work which has so far been described, and this is only skimming the surface of the overall picture to indicate the general areas of responsibility and some current issues, that two needs can be identified for the NRA. The first need is that of resources, the second is that of research.

On the first point the NRA has approximately 6,500 staff spread throughout England and Wales. It is currently setting up a network of 11 laboratories all of which will be fully operational by June 1991 and most of which will be in operation from early this year. Those laboratories currently employ over 200 staff but is expected that 100 new posts will be created when the network is complete.   Currently we process about 383,000 samples per year and anticipate that this could rise by 40% by next year. More importantly the number of measurements made on these samples will rise from 3,000,000 to very nearly 6,000,000 as a result of the introduction of automatic laboratory instrumentation capable of high sample throughput. The budget for capital equipment in laboratories will rise from £2M to approximately £4M by 1991.

In addition to laboratory work, scientific staff are involved in pollution control activities, in investigating incidents and advising on remedial action, setting standards on discharge for water quality and the maintenance of Water Act registers. This will require the development of very substantial computerised data systems and these are also being planned at the present time. It is clear from the foregoing that the problems with which the NRA is confronted are enormous.   The public, increasingly concerned about environmental matters, expect solutions to be found to insoluble problems.

In the second case the Water Act imposes on us to make arrangements for carrying out research. The NRA's current Research and Development Programme consists primarily of inherited projects previously carried out by the Water Authorities or by the Water Research Centre, together with a number of projects funded by the Water Directorate of the DoE. The spend arising from these programmes in 1989/90 is £6.4M, comprising 70% on applied research, with the remainder on experimental development. We have carried out a major review of all existing programmes together with substantial consultation with external organisations. More than 400 proposals have been considered in identifying 120 high priority new projects and it is planned that the R&D programme will increase over the next few years to an expenditure level of about £13M by 1992/93. Part of this programme is related to the development of automatic sampling and analysis techniques so that we are no longer in the position of relying on a limited number of samples taken at moments which do not necessarily coincide with pollution events. The use of computer modelling of pollution pathways and dispersion processes and broad environmental quality assessment techniques such as ecotoxicology and the use of biological indicators are also high on the list of priorities.

INTEGRATED POLLULTION CONTROL

The complications inherent in dealing with current environmental problems may be exacerbated to some extent by a proposed change in legislation through the Government's Environmental Protection Bill which is currently on passage through Parliament. In this the concept of integrated pollution control is set out for processes producing particularly dangerous substances, which will be known as "prescribed processes". The basis for IPC is the objective of controlling the discharges of such substances to land, water and air, and such processes will be authorised under a single system of control. The Bill removes from the NRA the responsibility for a large number of industrial processes and for probably 80% of all industrial discharges. The NRA whilst fully supporting the concept regrets that the system is being introduced in such a way as to take away these responsibilities so recently given it by the Water Act. The NRA is thus having vigorous exchanges with the Department of the Environment to ensure that the return of responsibilities to Her Majesty's Inspectorate of Pollution (HMIP) will not reduce the effectiveness of pollution control in the water environment. The problems of dealing with mixed sites where prescribed and non-prescribed processes co-exist must be resolved and at present the NRA is in the process of drawing up a Memorandum of Understanding between itself and HMIP in order to resolve this difficulty.

CONCLUSION

The NRA has been set up with a very wide range of duties and responsibility as set out in the Water Act 1989. Duties cover all aspects of aquatic environmental management but this paper has concerned itself solely with quality matters.

The NRA as presently constituted is the largest environmental protection agency in Europe, and will remain so despite the introduction of integrated pollution control. The NRA is an independent agency outside of Government and absolutely determined to carry out the duties that it has been given, and in this respect has already demonstrated its ability and determination to carry out this role thereby gaining widespread public support. We intend to be firm but fair with polluters and discharges and to achieve the objectives of improving and maintaining water quality and to this end regard ourselves as the guardians of the water environment.

Chapter 2

EUROPEAN DRINKING WATER STANDARDS AND THEIR IMPLICATIONS

M-M Bourbigots (Compagnie Générale des Eaux, Maisons-Laffitte, France)

The EEC directive has given rise to a set of national regulations published in the countries.

It represents a profound change in philosophy : the former rules, based on potability criteria have been replaced by rules based on quality objectives. As we shall see later on, when difficulties occur over respecting the standards, particularly with regard to pesticides, discussions are often focused on the toxic nature of these products, whereas the real debate concerns the technical and economic problems created by the application of these quality criteria, and the time required before strict conformity can be reached : **the debate is a political one.**

The distinction is not understood by public opinion and a great effort is required to popularize the debat. In difficult circumstances, for instance, in the event of accidental pollution or drought and so on, temporary waivers may be decided by the authorities with regards to some of the qualities fixed by the regulations. Public opinion often wrongly jumps to the conclusion that the water is not drinkable. Neither do some people understand why water that is not considered to be potable in one country may be judged drinkable in another (see Table 1).

In France, the water supplier is answerable to his customers for the constant respect of these objectives. He is free to adapt the control routine of his laboratory as he thinks fit. Moreover, he is also controlled by a state laboratory appointed by the relevant department of the Ministry of Health at intervals and according to a programme laid down in the regulations. The water supplies has to bear the financial burden of this control.

Inquiries were conducted by the Ministry of Health, the Basin Authorities and the private and public water suppliers. For most of the parameters except Nitrates and pesticides, the pollution are very localized and their excess can be dealt with additional treatment and structural renovations. A far more worrying aspect is agricultural pollution caused by nitrates and pesticides.

NITRATES

1 - ORIGIN

The importance of the nitrogen input to cultivated soil shows that agriculture, in the broadest sense of the term (including animal husbandry) is one of the main potential causes of water quality determination by nitrates, seen as nationwide problem. On a local scale, however, things may work out differently. Some sources, such as these generated by household effluents can reach quite large proportions compared with farming activities. However, remedial solutions can be found to alleviate this kind of pollution.

The situation when it comes to agricultural pollution is quite another matter. Its main characteristic is its wide diffusion making the struggle to protect the quality of water highly complex.

© 1991 Elsevier Science Publishers Ltd, England
Water Treatment – Proceedings of the 1st International Conference, pp. 17–22

Parameters	US National Primary Drinking Water Regulations Maximum contaminant level (MCL)		Europe Directive 80-778 July 15, 1980
	current	proposed	
Chromium	0.05 mg/l	May 89 0.10 mg/l	0.05 mg/l
Mercury	0.02 mg/l	May 89 0.002 mg/l	0.001 mg/l
Selenium	0.01 mg/l	May 89 0.05 mg/l	0.01 mg/l
Nitrate (as N)	10 mg/l	May 89 10 mg/l	50 mg/l (as NO_3) 11 mg/l (as N)
Nitrite (as N)		May 89 1 mg/l	0.1 mg/l (as NO_2) 0.03 mg/l (as N)
Atrazine		July 89 0.003 mg/l	0.0001 mg/l
PCB_s		May 89 0.0005 mg/l	0.0005 mg/l

Table 1 : Comparison of regulations in US and in Europe.
Some examples of differencies in MCL

A dozen or so sounding were carried out on the entire depth of the ground between the top soil and the phreatic layer in different regions by the Geological and Mining Research Bureau in France (BRGM) (1). Their results bring interesting elements :

On all cultivated land a stock of nitrates has been detected migrating towards the aquifers, and this without any noticeable attenuation of its concentration at increasing depths. These results show that the possibilities of natural denitrification are limited. Just few cases of denitrification have been identified normaly in the chalk aquifer in northern France.

The nitrate concentrations in intersticial waters are rarely under 45 mg/l and the project carried out in Brittany under market gardening sites shows that the water contained in the non-saturated zone can reach a nitrate content in the region of 170 mg/l with an average of 130 mg/l.

In intensely cultivated areas, the total quantities of nitrogen involved per annum are in the region of 200 to 250 kg per hectare (Nitrogen of the soil plus extra input from fertilizers). It can be assumed that the annual range of residual nitrates likely to be leached into the subsoil is between 30 and 45 kg of nitrogen per hectare (15 % of the stock in soil). This leakage produce concentrations in the aquifers of dose on 50 to 60 mg/l of NO_3^-.

This indicates that in order to maintain the quality of the groundwater, the farming practice should be able to reduce the leakage of nitrogen under 15 % of the total stock of nitrogen contained in the soil. This is a difficult challenge.

The origin of the increase in nitrate concentration doses not appear to be solely due to an increase in the doses of nitrous fertilizers, but just as much to the changes in crop rotation and above all a great increase in cropped areas. The draft Directive of the Council of the European Communities dated 12/01/1989 is intended to define a number of processes designed to protect the water against nitrate pollution. Member state would have to define all the areas in which water is exposed to the risk of pollution by nitrogen compounds. They would have to take all the measures required in order to respect a number of rules concerning the spreading of nitrogen compounds on soil, soil managment practices, as well as the treatment of municipal waste water.

The Ministers of Agriculture and Environment in France have set up a nationwide systems for the coordination of actions with the mission of developing a practical program called CORPEN (Orientation Committee for the reduction of nitrate-induced water pollution). It groups representatives of the agricultural world, fertilizer manufactures, professional institutions, users, research organisations, basin authorities, general government departments and qualified people.

2 - REGULATION

At the end of 1985, it was found that in several French Departments, despite the efforts made since 1981, the european standard of 50 mg/l could not be observed within the deadline fixed by the EEC (August 1985). The Minister of Health decided to introduce waiver proceedings, provided the following conditions were respected :

- the measure must be exceptional,
- only determined supply systems must be concerned,
- a proper program of improvements to the situation must be drawn up on the one hand and health officers and consumers must be kept informed on the other.

There can be no waiver in favour of :

- water from new catchments,
- packaged waters.

The waiver is granted in the name of the State by the Prefect of the department under a by-law issued by the French equivalent of the county council and published in an official gazette of the department concerned. Copies must be posted in Town Halls and sent to the Minister of Health.

3 - SOLUTIONS

Numérous remedial actions were of course implemented since 1981, including the following changes :

- changes in pumping levels,
- the abandon of certain wells,
- the research on new ressources,
- interconnections with other supply networks,
- investments in denitrification process.

The four first solutions are the most sought.

Over 90 % of the water distribution units showing concentration of more than 50 mg/l are situated in rural districts. Over 95 % of the distribution units affected supply under 5,000 inhabitants, and over 70 % supply less than 1,000 (2).

One can understand that for many of these small units, the investments to improve the water quality are difficult to apply owing to the very meagre financial resources.

Specific nitrate removal treatment must be required when alternative solutions are not feasible.

The technical solutions are numerous. For example we can mention (3, 4, 5, 6) :

- process using membranes : reverse osmosis, electrodyalisis,

- ion exchange processes : permutation of nitrate ions by chloride and/or bicarbonate ions,

- biological processes : heterotrophic with addition of inorganic carbon or by using hydrogen as a reducing reagent.

Research started in 1978 led to the construction in 1983 of the first heterotrophic biological facility in the world applied on an industrial scale at Eragny-sur-Oise (France).

Since 1983, several plants were built using the ion exchange or biological processes (see Table 2).

The biological method still remains the only one that entirely solves the problem by transforming the nitrates into elementary nitrogen restored to the atmosphere. However, the search for the best cost ratio for this type of installation has led to propose the ion exchange method as an economically attractive alternative for small production rates, despite the ecological problem created by the discharge of salty effluents inherent in the process.

Biological denitrification		
Dennemont (Yvelines)	400 m³/h	OTV
Eragny (Val d'Oise)	80 m³/h	OTV
Champfleur (Sarthe)	70 m³/h	Degremont
Château Landon	50 m³/h	Degremont
Ion-exchange resins		
Binic (Côtes d'Armor)	160 m³/h	OTV
Craon (Mayenne)	15 m³/h	OTV
Heiltz l'Evêque (Marne)	13 m³/h	OTV
Semalens (Tarn)	36 m³/h	OTV
Nicolas de la Grave (Tarn et Garonne)	65 m³/h	OTV
St Adresse (Seine Maritime)	150 m³/h	OTV
Crépy en Valois (Oise)	25 m³/h	SAUR
Ormes sur Voulzie (Seine et Marne)	30 m³/h	Degremont
Plouenan (Finistère)	500 m³/h	Degremont
Plounevez-Lochrist (Finistère)	50 m³/h	SAUR

Table 2 : List of plant using the biological denitrification
or the ion-exchange resins processes in France (non exhaustive list)

PESTICIDES

1 - REGULATION

The word pesticides is a very loose one and gathers under one hat all the chemicals used for the destruction of organims. These products may be organic or inorganic.

As far as organic molecules are concerned, they are of different structures, polarity, size, ionic form, volatility, etc. It therefore seems quite impossible to determine a pesticide index. Analysing the pesticides one by one, appears physically very difficult. In France, for instance, over three hundred active substances are used in agriculture in formulations numbered in thousands.

The EEC Directive 80/778 tells us that "by pesticide and related product is meant :

- insecticides,
- herbicides,
- fungicides,
- PCB and PCT.

It must be remembered that drafting of the directive 80/778 began in 1974/75 at a time when the only compounds investigated were organo-chlorinated and organo-phosporous pesticides, and often for the latter only these that respond to electron capture, which meant that it was possible to quantify all the compounds present with a single analysis.

The maximum admissible concentration level of 0.1 µg/l corresponds to the detection limit in the years 70 to 75 and it must be stated that, in the law makers, pesticides correspond only to organo-chlorine and/or phosporous compounds, therefore these thresholds represent the toxicity of these compounds.

The difficulties met to observe the maximum concentration level for the herbicides, specially for the triazines and the lengthy debate on the subject illustrates the part played by political factors in determining standards. It is obvious that the standard adopted for the triazines and the other pesticides cannot be the result of a strictly scientific approach. We are going to find ourselves increasingly confronted with decisions made by politicians on matters for which we hold no scientific evidence.

It is not surprising therefore that difficulties in applying such standards stir up the true debate, the political one : who can and should take action ? who should pay ? what is the priority action to be taken ?

With a view to protecting the quality of drinking water, regulations have ben drawn up for the protection of groundwater and surface water catchment sites. They include the installation of different protective perimeters round the intake areas. However, these measures are not always easy to implement. The regulation must above all be backed up by a campaign of information and widespread advisory and consultancy services to users of these plant protective products.

There is a question to be answered, however : have we reached contradictory constraints, the protection of water and agriculture, by imposing such a low level ? We hope that the answer is no and that we can find the solution.

2 - TREATMENT

Water is largely contaminated by atrazine and simazine. Other pesticides appear at the trace level, but it is also important to study these that are likely to predominate on the market in the coming years.

Treatments have been tested on atrazine and simazine. Adsorption on granular or powdered activated carbon and ozonation seems to be the more efficient treatments. The use of ozonation often explains the high removal rates observed on treatment train. The efficiency achieved at the outlet of the ozonation step varies between 25 and 50 % depending on the dose rates and the contact time.

The efficiency of activates carbon is highly dependent on its degree of saturation, contact time and granulometry. In the case where Picabiol type carbon is used, with 10 minutes contact time, for treating an atrazine concentration of about 0.5 µg/l, the Directive will be satisfied for a year.

In conclusion, the efficiency of a complete conventional treatment train with an ultimate refining step, using a combination of O_3 and GAC, varies between 60 and100 % removal depending on the degree of which the carbon is saturated.

Ozonation combined with hydrogen peroxide or UV seems to be more effective that ozone alone.

In a case where the TOC concentration is 2 mg/l, alcalinity 20.8 °F and the ozonation rate 2.5 mg/l, an H_2O_2/O_3 ratio of 0.4 g/g provides 90 to 95 % abatement of atrazine (7, 8, 9, 10). This oxidation is now applied in France on the industrial scale on groundwater and on

Concentration of atrazine in raw water		0.5	1.5
Treatment with PAC		0.35 FF/m^3	0.60 FF/m^3
Treatment with GAC (including regeneration)		0.15 FF/m^3	0.20 FF/m^3
extra cost when ozonation is already applied in the treatment train	GAC (including regeneration)	0.15 FF/m^3	0.15 FF/m^3
	Hydrogen peroxide	0.02 FF/m^3	0.03 FF/m^3

Table 3 : Extra operating costs required to respect the standard of 0.1 µg/l
of atrazine in the water

two surface waters. Concerning the surface waters, a temporary authorization has been given by the Health Minister. During the first year of application, specific control will be done by a state laboratory appointed by the relevant department of the Ministry of Health.

The extra operating cost required in order to respect the standard of 0.1 µg/l of atrazine in the water is dependent on the treatment train and the concentration of atrazine in raw water (table 3).

The necessary level of investments is so dependent on the size of the plant that it is impossible to generalize.

CONCLUSION

These new rules in water treatment show the difficulty created by a change in the regulation.

The wording of the standard should be explained more clearly together with analytical methods and how the results must be interpreted.

The water supplier is answerable to his customers for the constant respect of the potability and a great effort must be required to popularize the debate and the difficulties occur over respecting the new standards.

REFERENCES

(1)   Landreau, A. Les nitrates dans les eaux souterraines. COURANTS, Revue de l'Eau et de l'Aménagement, 1, 1990.

(2)   Ministère chargé de la Santé, Direction générale de la Santé. Teneurs en nitrates des eaux destinées à la consommation humaine en 1985-1986-1987.

(3)   Philipot, J.M. and de Larminat, G. Nitrate removal by ion exchange : the Ecodenit process, an industrial scale facility at Binic (France). Wat. Supply, 6 : 45-50, 1988.

(4)   Ravarini, P., Coutelle, J. and Damez, F. L'usine de Dennemont. Une unité de dénitrification-nitrification à grande échelle. TSM-l'Eau, 4 : 235-240, 1988.

(5)   Dries, D., Liessens, J. Verstraete, W. et al. Nitrate removal from drinking water by means of hydrogenotrophic denitrifiers in a polyurethane carrier reactor. Wat. Supply, 6 : 181-192.

(6)   Liessens, J., Germonpré, R. and Vertraete, W. Comparative study of processes for the biological denitrification of drinking water. Proceedings Forum fo Applied Biotechnology, State University of Gent, Belgium, September 28, 1989.

(7)   Paillard, H., Partington J., Valentis, G. Technologies available to upgrade potable waterworks for triazines removal. IWEM Conferences on pesticides in the environment, London, 12 avril 1990.

(8)   Paillard, H., Legube, B., Gibert, M. and Dore, M. Removal of nitrogenous pesticides by direct and radical type ozonation. Proceeding of the 6th European Symposium on organic micropollutants in the aquatic environment, Lisbon, 22-24 May 1990.

(9)   Paillard, H. Valentis, G., Partington, J. and Tanghe, N. Removal of atrazine and simazine by O3/H2O2 oxidation in potable water treatment. Proceedings of 1990. IOA.PAC Conference, Shreveport, Louisiana, US, 1990.

(10)   Paillard, H. Traitement des micropolluants organiques. Conférence aux Journées de Choisy, 21 juin 1990.

Chapter 3

COST ESTIMATION OF DIFFERENT AMBIENT WATER QUALITY PROTECTION POLICIES

A Filip and D Obradović (Energoprojekt Holding Corporation, Belgrade, Yugoslavia)

ABSTRACT

A mathematical model describing the ambient water quality in a part of the West Morava watershed is used for evaluating different water quality protection policies. For anticipated degrees of ambient water quality protection, the necessary decrease of pollution emission for 1990 is evaluated and costs are estimated.

INTRODUCTION

Beneficial uses of ambient waters in a part of the West Morava river basin are defined by Federal Regulations in the following order: public water supply, fishery, irrigation and recreation. All ambient waters are classified and strict water quality criteria are set. But, due to rapidly increasing population migration towards river valleys, pollution loads have been also considerably increased in recent years and water quality standards are in some places violated, wholly or partially (too high carbonaceous or /and nitrogenous BOD, low content of dissolved oxygen, too many coliforms in water, etc.).

This report presents an approach for solving the problem of serious ambient water quality deterioration in a part of the West Morava river basin with a drainage area of 1250 sq. kilometers and the total river length of 94 kilometers.

The mathematical model of water quality in the West Morava, based upon the program STEKAR (Gherini and Summers, 1978), was made and calibrated. Different alternatives for waste water treatment (point sources only) were proposed - each one consisting of a set of prescribed degrees of waste water treatment for each point source (Filip and Obradović, 1984). The effects of alternative solutions were then analysed on the mathematical model for the low flow conditions.

FEATURES OF THE STUDY AREA

Basin Description
The area of the studied part of the basin is about 1250 sq. kilometers. The elevation of the highest point is 330 metres, that of the lowest point being at 190 metres above sea level. The precipitation ranges from 700 to 1100 mm per year. The wettest period of the year is late spring-early summer; the driest period is winter. The total length of the river course covered in this part of the study is approximately 94 kilometers (see Fig.1).

River Classification and Ambient Water Quality Standards
Ambient water quality classification is given in Table 1. Classes I, II-a and II-b respectively correspond to categories A_1, A_2 and A_3 as defined by the Council of European Communities (Luxemburg, June 1965).

Water Treatment – Proceedings of the 1st International Conference, pp. 23–30

Table 1. Classification of Ambient Waters

River Class	Class Limiting Parameter	Criteria* Value	Water Use
I	Dissolved Oxygen	8	. Drinking Water
	BOD_5	2	Supply (disinfection only)
	TDS	350	. Fishery
	Coliforms	200	. Recreation
II A	Dissolved Oxygen	6	. Drinking Water
	BOD_5	4	Supply (conventional treatment)
	TDS	1000	. Fishery
	Coliforms	6000	. Recreation
II B	Dissolved Oxygen	5	. Substantially the
	BOD_5	6	same as for II A
	TDS	1000	
	Coliforms	6000	
III	Dissolved Oxygen	4	. Industrial Water
	BOD_5	7	Supply
	TDS	1500	
	Coliforms	10000	
IV	Dissolved Oxygen	0.5	. Out of use
	BOD_5	unlimited	
	Coliforms	unlimited	

* All values in mg/l. Coliforms in MPN/100 ml

Pollution Sources
Two major towns - Čačak and Gornji Milanovac are situated in this area. Both are important industrial centres. The pollution loads from these towns, municipal and industrial, expected in 1990 are listed in Table 2. These data are based partly on extensive field measurements of waste water discharges in a twelwe-month period in 1978-1979 and partly upon regional development plans and enquiries in industry. Non-point sources of pollution were neglected as low flow condition are simulated.

MODEL CALIBRATION

Description of the Program STEKAR
The river system is divided in to) segments with homogeneous characteristics and represented as nodes in the mathematical model. For each node, the following data are collected: length, average depth, surface area, volume. The nodes are linked by channels, each one being characterized by: cross-section area, depth, velocity of flow, bottom slope. The data on ambient water quality, weather conditions and waste water inputs were also necessary. The node network and locations of waste water inputs are shown in Fig. 2.

Calibration of the Model
Extensive field measurements of ambient water quality were conducted simultaneously with the measurement of point source emissions. The same situations are then simulated by using adequate boundary and initial conditions. Reaeration constants, decay rates and other constants were determined by this analysis. The constants for low-flow conditions were evaluated on the basis of similar hydrological and weather situations.

Waste Water Treatment Levels
Standardized waste water treatment plants for towns have been supposed and unit

Tabele 2: Point Sources Pollution Loads in 1990.

S o u r c e	Discharge lps	Ultimate C-BOD mg/l	Ultimate N-BOD mg/l	Suspended Solids mg/l	Dry Re- sidue mg/l
GORNJI MILANOVAC					
. municipal	97	468	47	210	950
. industrial	16	9	2	67	1200
ČAČAK					
. municipal	346	348	43	250	550
. industrial	443	8	2,5	420	550
TRBUŠANI					
. industrial	97	8	2,5	5195	330

costs calculated. These plants can be upgraded and enlarged if such a necessity arises later. The accepted treatment efficiency for sucha plant can be found in Table 3. For industrial wastes that are not compatible with the municipal ones, separate physico-chemical PHC treatments or pretreatments are foreseen.

Table 3: Waste Water Treatment Levels

Treatment	Removal, percentage		
	BOD	Susp. solids	Ammonia
Primary (P)	30	60	0
Secondary (S)	85	85	60
Tertiary (T)	96	96	96

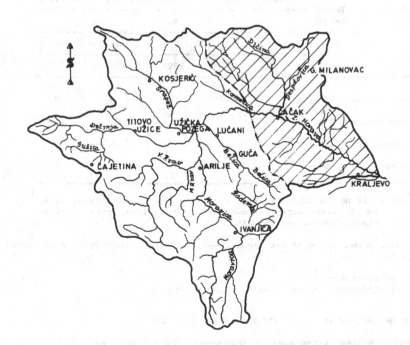

Fig.1. West Morava river basin with location of the studied area

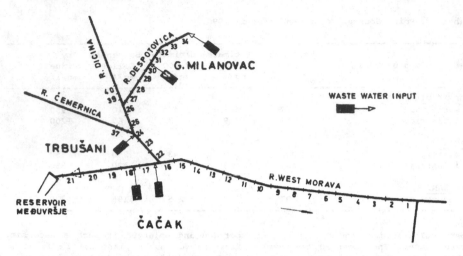

Fig.2. Layout of the mathematical model for simulation of water
        quality in the part of the West Morava river basin

VIOLATIONS OF AMBIENT WATER QUALITY STANDARDS IN THE STUDY AREA

Violation of ambient water quality standards for pollution loads as measured in
1978/79 and predicted pollution loads in 1990 are presented in Table 4.

Table 4. Extent of Ambient Water Quality Standards Violation

River	Total Length km	Length of the River Reach Where WQ is Violated			
		km			
		BOD$_5$		Dissolved Oxygen	
		1978	1990	1978/79	1990
West Morava	61	61	61	0	0
Despotovica	15	0*	0*	10	10
Dičina	12	9	9	9	9
Čemernica	6	6	6	6	6

* There is no limitation of BOD$_5$ concentration in Despotovica river
  - from the waste water inflows down to the Dičina river, due to
  its classification in Class IV.

The calibrated model was used for simulation of ambient water quality in low-flow
conditions for cases:

- no waste water treatment,
- effects of different degrees of waste water treatment.

DEFINITION OF AMBIENT WATER QUALITY PROTECTION POLICIES

The step-wise ambient water quality improvement is accepted as a policy, so the
costs will be more easily met.

Ambient water quality goals are listed in Table 5, given as maximum permissible concentration (MPC). Complete removal of floating debris, grease and oil as well as substantial removal of coliforms are anticipated. Implementation of the third degree of protection ensures the class of stream prescribed by the Regulations. Water quality of this stream is wholly protected. If the fourth degree of protection is introduced, some of the river reaches may attain a higher class.

Table 5. Ambient Water Quality Improvement Goals

Degree of Protection	Prevention of MPC Violation				Improvement of Stream Category
	Toxic Materials and Coliforms	Dissolved Oxygen	Suspended Solids	BOD_5	
First	+				
Second	+	+			
Third	+	+	+	+	
Fourth	+	+	+	+	+

BENEFITS OF IMPLEMENTATION OF CERTAIN DEGREES OF PROTECTION

The benefits to be gained from this step-wise protection of deteriorated ambient water quality are as follows:

First Degree of Protection

Benefits of the applied first degree of protection are:

- water may be used for watering of cattle and irrigation;

- WQ in downstream reaches will be improved, with a corresponding benefit to local users;

- aquatic population is protected;

- aesthetic appearance of the stream and environment will be improved;

- better situation in the water supply system located further downstream, either in the case of direct intake, or in the case of infiltration basins/ground water aquifer use.

Second Degree of Protection

Benefits of the applied second degree of protection are:

- all positive effects of the first degree are included plus the following benefits:
  . further improvement of water quality for all uses;
  . prevention of septic conditions in some river reaches;
  . better self-purification characteristics of the stream;
  . full protection of aquatic life;
  . possibility of using the river for recreation.

Third Degree of Protection

Benefits of the applied third degree of protection are:

- all positive effects of both previous degrees of protection are included plus the following benefits:

  . return of the stream in its class as prescribed by the Regulations;

- guaranteed possibility of using river for water supply of towns and industry with usual treatment only;

- better organoleptic characteristics of water;

- reduction of water quality deterioration in the distribution network.

Fourth Degree of Protection

Benefits of the applied fourth degree of protection are:

- all positive effects of all three previous degrees of protection are included plus the following benefits:

  - improvement of water quality in the Despotovica river, from Gornji Milanovac to Dičina, enough to raise it to a higher class. Water quality in the Dičina, the Čemernica and the West Morava will be then improved to such a degree that reclassification of these rivers could be achieved.

NECESSARY DEGREE OF WASTE LOADS REDUCTION AND COST ESTIMATION

The desired degree of protection can be achieved only by applying the definite level of waste water treatment. Standardized waste water treatment plants for towns

Table 6. Necessary Degree of Treatment and Costs of Protection for the 1990 Pollution

Degree of Protection	Pollution Source	Treatment Level	Treatment Costs $(10^6$ din.) [*]	
			Investment Costs	Annual Costs
First Degree of Protection	Čačak II	P	45.70	9.88
	Industry	PHC	8.80	2.13
	Paper Mill	PHC	8.36	2.37
	Trbušani	PHC	11.19	2.13
	G. Milanovac	P	40.15	9.27
	T o t a l:		114.20	25.78
Second Degree of Protection	Čačak II	S	80.15	18.20
	Industry	PHC	8.80	2.13
	Paper Mill	PHC	8.36	2.37
	Trbušani	PHC	11.19	2.13
	G. Milanovac	S	59.85	15.72
	T o t a l:		168.35	40.55
Third Degree of Protection	Čačak II	T	104.20	23.66
	Industry	PHC	8.80	2.13
	Paper Mill	PHC	8.36	2.37
	Trbušani	PHC	11.19	2.13
	G. Milanovac	S	59.85	15.72
	T o t a l:		192.40	46.01
Fourth Degree of Protection	Čačak II	T	104.20	23.66
	Industry	PHC	8.80	2.13
	Paper Mill	PHC	8.36	2.37
	Trbušani	T	14.54	2.78
	G. Milanovac	T	77.80	20.43
	T o t a l:		213.70	51.37

[*] 1 DIN = 0.08 US$

have been supposed and unit costs calculated. These plants can be upgraded and enlarged if such a necessity arises later. For industrial wastes that are not compatible with the municipal ones, separate physico-chemical treatments or pretreatments are foreseen. Several alternatives with a different degree of waste water treatment have been simulated on a computer (Filip, Obradović, 1984), and feasible solutions for ambient water quality protection in the region were elaborated.

Investment costs were based on the 1990 prices level. Annual costs consist of annuities for civil works based on 13.6 percent of investment, annuities for equipment based on 16.2 percent of equipment costs as well as operational and maintenance costs.

The necessary degree of waste loads treatment as well as investment and annual costs for the 1990 pollution are presented in Table 6. Cost estimation for protection measures for the 1990 pollution are based on the assumption that no actions nave been taken before 1990.

SUMMARY AND CONCLUSIONS

The mathematical model describing the ambient water quality in West Morava watershed are used for evaluating different water quality protection policies. For anticipated degrees of ambient water quality protection the necessary decrease of pollution emission for 1990 is evaluated and costs are estimated.

ACKONOWLEDGEMENTS

The Water Resources Authorities of Serbia and the International Bank for Reconstruction and Development have provided the funds for the study "Protection of Water Quality in the West Morava" from which a part is presented here.

The Authors are acknowledging with gratitude the fine cooperation of Mr. Carl Chen, Ph.D., Mr. Steve Gherini, Ph. D., Mr. James Kelly and Mr. William Mills, all from Tetra Tech, Inc., Laffayette, California, for consultations and help extended in the calibration of this mathematical model. The Authors are also grateful to the staff of Energoprojekt Co. and Institute Jaroslav Černi, Belgrade, that have participated in waste water measurements, data collection and waste treatment process development in the framework of this study.

REFERENCES

1. Gherini, S. and Summers, K. (1968): River Water Quality Model for the Northern Great Plains", User's Manual for the program STEKAR, TETRA TECH Co., TC-668.

2. Water Quality in the Western Morava Region. Appendix E-Alternative Plans and Programs. Energoprojekt, Jaroslav Černi Institute, Tetra Tech Inc. Belgrade 1980.

3. Filip, A. and Obradović, D., Evaluation of Alternative Measures for Controlling Pollution in a Part of the West Morava River Basin. Paper Elaborated for IAWPRC Conference, Amsterdam 1984.

Chapter 4

WATER SUPPLY QUALITY AND ITS REGULATION IN THE PRIVATISED UK WATER SERVICES plcs

S B Tuckwell (Wessex Water Business Services Ltd, UK)

1 ABSTRACT

Water supplies in England and Wales have to comply with revised statutory quality and monitoring requirements, incorporating the EC Drinking Water Directive values and national standards. Despite up to 99.7% of samples meeting the standards, extensive programmes are required to improve supplies to try to achieve 100% compliance, particularly for water treatment works to remove aluminium, iron and pesticide and for distribution systems for iron and related problems due to corrosion of mains. A more stringent standard for lead than in the EC Directive will require extensive investigation of possible non-compliance due to domestic plumbing materials and may necessitate water treatment remedies. The £10 billion improvement programme should remedy most of the problems by 1999 or sooner.

2 STRUCTURE OF THE NEW INDUSTRY

The Water Act 1989 (HMSO 1989a) prepared the way for the privatisation of the ten regional water authorities in England and Wales which, since 1974, had been responsible for water supply and sewerage services, resource planning, river water quality, pollution control, fisheries, land drainage, flood protection and alleviation, water recreation and conservation.

Fig. 1: England and Wales showing the ten Water Services companies.

Water Treatment – Proceedings of the 1st International Conference, pp. 31–42

The Act enabled ten Water Holding Companies to be formed, each of which holds a Water Services Company and other subsidiary companies. These Water Service Companies (Fig. 1) have been granted twentyfive year licences to provide water supply (except in areas served by the already existing statutory water companies) and sewerage services in geographically defined areas of the former water authorities. The assets associated with water supply, sewerage and sewage treatment were transferred to the Water Service Companies. The majority of the remaining duties of the former water authorities became the responsibility of a new public body, the National Rivers Authority.

Each Water Service Company is under a duty to develop and maintain an efficient and economical water supply system in its water region. It has similar duties for sewage collection and disposal in the Sewerage Region.

3 REGULATORY PROVISIONS

The activities of the Water Service Companies are regulated by a series of new and existing controls and legislation. The principal provisions are set out in the Water Act, in regulations made under the Act and in conditions contained in the operating licences. Activities which are not related to the "core" business of water and sewage treatment are generally not controlled by these regulations. The principal regulators are the Director General of the Office of Water Services, the National Rivers Authority and the Secretary of State for the Environment.

3.1 Duties Of The Director General

The primary duties of the Director General are to ensure that the functions of the water and sewage undertakers are properly carried out - with special respect to protecting customers, facilitating competition and promoting economy and efficiency. To compensate for the lack of direct competition for services in any company area, the Director General will use regulation and inter-company comparison of performance. The most important way in which regulation is to be used is in the control of the charges which companies are allowed to make for water services. The Director General has set limits on charges to ensure value for money, efficiency and adequate investment for meeting current and future needs of customers for water services. He ensures there is no unfair discrimination between different groups of customers, for example between the charges for metered and unmetered domestic water supply users. He has set up Customer Service Committees for each main region, which have the duty to ensure that companies deal properly with customers' queries, complaints and other interests. Each company has to publish a Code of Practice for customers, setting out its services, procedures for complaints and emergencies and its charges and methods of payment. There must be other codes on matters such as disconnections of supplies and leakage of water from metered and unmetered properties. Another novel requirement is the Guaranteed Standards Scheme, whereby customers are entitled to monetary compensation if the company breaches certain standards of service for interruption of supplies, response to billing or payment enquiries, written complaints about water supply or sewerage services and failures to keep appointments.

3.2 Duties Of The National Rivers Authority (NRA).

In relation to water supplies, the NRA has the following duties:

> to control water pollution

> to maintain and improve the quality of waters used for water supply

> to operate the water resource abstraction licensing system

> to ensure the proper management of water resources and to assess the adequacy of water resources against future needs

> to assess the quality of surface waters intended for abstraction of drinking water, under the terms of the EC Directive (75/440/EEC).

There are other duties relating to the promotion of nature conservation and recreational use of waters.

3.3 Duties Of The Secretary of State.

The Water Act gives many powers to the Secretary of State. Of particular relevance is the power to make regulations to govern the quality of drinking water.

4 WATER SUPPLY QUALITY REGULATIONS

The Water Supply (Water Quality) Regulations (HMSO 1989b,c), approved by Parliament in July 1989, came into force in September 1989, (except for the monitoring provisions, in force from 1st January 1990). They cover water treatment, standards of water supply quality, monitoring frequencies, action if water supplies fail to meet the standards, the reporting of results and the provision of information.

4.1 Water Treatment Requirements.

Except for certain groundwaters exempted by the Secretary of State, all water provided for drinking, cooking and washing must be disinfected. A certain minimum standard of treatment is specified for surface-derived waters to satisfy the requirements of the EC surface water directive (75/440/EEC) (EC 1975). The Secretary of State may require that only approved treatment processes be used or approved substances be added to water to treat or improve its quality and these approvals may have conditions attached which must be observed.

4.2 Wholesomeness Of Water Supplies.

Historically drinking water in the UK was required to be "wholesome" but there were no legal quality standards to define this. The EC directive (80/778/EEC) relating to the quality of water intended for human consumption (the Drinking Water Directive) (EC 1980) was considered by the UK Government to underline and reinforce existing UK policy and procedures and therefore it was implemented in the UK by government circular in 1982 (DoE 1982), with a deadline of July 1985 for compliance with its standards. It was the Government's view that compliance with the Directive was a necessary condition of wholesomeness, but not a complete definition of it. Following a challenge by the European Commission of the U.K. interpretation of the EC Directive, the Government used the Water Act to incorporate the Directive's requirements into national law by means of regulations made by the Secretary of State. The Water Act (Section 52) places a duty to supply water to domestic premises which is wholesome and makes it a criminal offence to supply water which is unfit for human consumption.

4.3 Water Quality Standards.

The Water Supply Regulations define wholesomeness as meeting all the prescribed standards in the Regulations and not containing anything which could be harmful to health, either on its own or in combination with other substances in the water. The Regulations provide legally defined quality standards for drinking water for the first time in England and Wales (Appendix).(They have also been extended to Scotland and are under consideration for Northern Ireland.) They incorporate the same maximum admissible concentrations (MACs) defined in the Drinking Water Directive for fortytwo parameters and the qualifying comments for two other parameters (total organic carbon and total bacterial counts.) In addition, a minimum value of 5.5 has been adopted for pH. The standard for lead (0.05 mg/l) is more stringent because no flushing of the tap is permitted before sampling and a litre sample volume must be taken. The Minimum Required Concentrations for total hardness and alkalinity, applicable to water which is artificially softened during water treatment, are also specified.

Standards have also been defined for other parameters, including those in the EC Directive which have a Guide Value only (Table I). The limits for benzo 3,4 pyrene and the chlorinated solvents are based on World Health Organisation Guideline Values ((W.H.O. 1984).

Parameter	Units	Maximum concentration or value
Conductivity	uS/cm	1500 at 20 ° C
Chloride	mg Cl /l	400
Calcium	mg Ca /l	250
Substances extractable in chloroform	mg/l dry residue	1
Boron	ug B /l	2000
Barium	ug Ba /l	1000
Benzo 3,4 pyrene	ng/l	10
Tetrachloromethane	ug/l	3
Trichloroethene	ug/l	30
Tetrachloroethene	ug/l	10
Trihalomethanes	ug/l	100

Table I: Additional standards for water for domestic purposes.

Some of the descriptive standards in the EC directive have been excluded from the Water Quality Regulations. Dissolved oxygen is not considered meaningful for routine measurement and free carbon dioxide is not meaningful for water at the tap (Miller, 1990). The standard for hydrogen sulphide ("undetectable organoleptically") is considered to be included in the taste and odour parameters. These standards apply to water samples taken from taps in the distribution system, but the same microbiological standards are prescribed for water leaving treatment works and in service reservoirs.

4.4 Monitoring frequencies.

Water leaving treatment works must be monitored for total and faecal coliforms and disinfectant residual at frequencies depending upon the average annual output of water from the works, ranging from weekly to daily samples. Service reservoirs are monitored weekly for the same microbiological parameters.

The majority of standards apply to water from the distribution system and the sampling of water is based on zones. A single source, or a blend of several sources, feeding a discrete area must be designated as a single zone (subject to a maximum zone population of 50,000.) Areas with water quality variations within a zone should be subdivided. The sampling points must be representative of the whole zone and, for most parameters, these may be either fixed points or randomly selected customers' taps. At least 50% of the microbiological samples and all the samples examined for copper, lead and zinc must be from randomly selected customers' taps.

The minimum frequencies of analysis for tap samples are generally slightly greater than those of the EC Drinking Water Directive (where it specifies frequencies.) Frequencies are not specified in the Regulations for faecal streptococci, sulphite-reducing Clostridia, substances extractable in chloroform, Kjeldahl nitrogen, dissolved hydrocarbons or phenols, although the Regulations specify standards for them. This is in line with the discretion allowed the 'competent national authority' in the EC Directive. However, the Regulations require water suppliers to analyse for any parameter which they believe may exceed the standard. If any of the non-microbiological parameters exceeds the standards, increased monitoring frequencies are specified for the zone for up to two years. If water quality in any zone is consistently within the standards, the frequency of monitoring can be reduced. In addition to this compliance monitoring, water suppliers are expected to carry out sampling for operational monitoring.

An example of standard monitoring frequencies is given (Table II) for selected groups of parameters for the smallest and the largest permitted zones:

Parameter groups	Frequencies for zones of given population (number of samples per year)	
	Population <500	Population 50,000
Conductivity, qualitative taste & odour	4	60
Turbidity, nitrate, nitrite, iron, aluminium, colour	4	10
Lead, copper, zinc, trihalomethanes, pesticide, P A Hydrocarbon	4	4
Chloride, sulphate, calcium, fluoride, toxic metals.	1	1
Microbiological	12	60

Table II: Standard monitoring frequencies for the smallest and largest permitted sampling zones.

5 COMPLIANCE AND ENFORCEMENT

5.1 Interpretation of compliance requirements.

The 1982 U.K. Government interpretation of the EC Directive (DoE 1982) considered that for most parameters the average of samples taken over a period (usually one year) should be compared with the MACs to assess compliance. This was challenged by the European commission and now, for all the standards which derive from MACs, the Regulations require every distribution tap sample to comply with the standard (100% compliance). As in the Directive, the exceptions to this are for total coliforms and sodium. For total coliforms the water complies if 95 % of the samples in any period meet the standard. The period will normally be one year, but if less than fifty samples have been taken from a zone in one year the results of the previous fifty samples are taken regardless of the period. For sodium, 80% of samples taken in the past three years must not exceed the standard for the water to comply.

For the parameters in Table I compliance is based on the average of results of analyses over twelve months are compared with the standards. (For trihalomethanes a rolling three month period is used.) For samples of water leaving treatment works and in service reservoirs, every sample must comply with the bacteriological parameters for the water to be wholesome.

5.2 Relaxations of the Standards.
Subject to there being no risk to public health, the Secretary of State may relax the standards for geographically defined areas where supplies fail to comply for specified reasons, as permitted by the EC Directive (Article 9). Where supplies are affected due to emergencies or exceptional weather, temporary relaxations can be granted; temporary or permanent relaxations can be approved due to the nature or structure of the ground. Conditions may be imposed to require improvements to water quality or monitoring.

5.3 Section 20 Undertakings Given to the Secretary of State.
The Secretary of State has a duty under the Water Act to enforce the standards of wholesome water. However, Section 20 of the Act allows him discretion in enforcement where a) the contraventions are trivial or b) he is satisfied that the water supplier has given, and is complying with, an undertaking to take all necessary steps to achieve compliance (a "Section 20 Undertaking"). These are the type of action programmes referred to in Article 20 of the EC Drinking Water Directive. The Secretary of State can still enforce compliance against a company despite an Undertaking, if there are grounds for doing so.

6 REVIEW OF CURRENT QUALITY

6.1 Level of compliance.
In November 1989 each water holding company summarised its water quality in detailed prospectuses (Schroder, 1989) which were independently vetted by auditors as part of the water privatisation share offer. A general view of the current water quality and the difficulties faced can be drawn from this source. The companies have tried to assess retrospectively the compliance of their supplies in 1988/9 with the quality standards in the Regulations. In some cases the location and manner of sampling was dissimilar from the Regulations' requirements, but the extent of sampling was considerable. Thames Water, for example reported 471,000 tests for 53 of the 57 parameters in the Regulations with only 627 failures (0.13%) overall. Wessex Water estimated 99,000 tests on 9,700 samples from distribution systems and only 289 failures, (a compliance rate of 99.7%).

Reported compliance for tap samples ranged from 88 - 94% for aluminium, iron, manganese and trihalomethane in Yorkshire, 95.5% for iron in North West and overall, for a wide range of parameters, from 98.2 to 99.8% for other companies. The Regulations require 100 % compliance for most parameters and notwithstanding these high compliance figures, it was acknowledged by the Government that this could not be achieved instantaneously. Hence it permitted Relaxations and Section 20 Undertakings, relating them to water sampling zones, although the remedial action could refer to specific water treatment works or service reservoirs as appropriate. Water supply quality deficiencies which affected only part of the zone still required the whole zone to be nominated.

7 RELAXATIONS

Relaxations covering an estimated 3% of the population of England and Wales were granted in November 1989 for sodium, dissolved hydrocarbons, sulphate, magnesium, potassium, colour, aluminium and manganese. These are subject to review at the end of 1994 or earlier, when they might be renewed. Temporary relaxations have been granted, subject to agreed improvement programmes designed to achieve compliance by given dates up to 1995, for colour, turbidity, pH, odour, taste, oxidisability, nitrite, ammonium, aluminium, iron and manganese. The parameters covered by relaxations affecting the Water Services companies are given in Table III.

PARAMETER	ANGLIAN	NORTH-UMBRIAN	NORTH WEST	SEVERN TRENT	SOUTH-ERN	SOUTH WEST	THAMES	WELSH	WESSEX	YORKSHIRE
Aluminium		*	*	*		*		*		*
Ammonia	*						*			
Colour	*	*	*			*		*		*
Iron	*	*	*	*	*	*	*	*	*	*
Magnesium				*						
Manganese	*	*	*	*	*	*		*	*	*
Nitrite							*			
Odour	*			*	*	*		*		
Oxidisability								*		
pH								*		
Potassium									*	
Sodium				*						
Sulphate				*						
Taste	*			*	*			*		
Turbidity	*		*	*		*	*	*	*	*

Table III: Parameters for which Relaxations have been granted to Water Services companies

PARAMETER	ANGLIAN	NORTH-UMBRIAN	NORTH WEST	SEVERN TRENT	SOUTH-ERN	SOUTH WEST	THAMES	WELSH	WESSEX	YORKSHIRE
Aluminium			*	*	*	*	*	*	*	*
Ammonia	*									
Cadmium			*							
Fluoride	*									
Iron (at works)	*		*					*		*
Manganese								*	*	
Nitrate	*			*			*			*
Nitrite	*		*				*			*
o-Phosphate	*									
PAHydrocarbon			*		*	*	*	*		
Pesticide	*	?	?	*	*		*	*	*	
pH								*		
THM		*	*	*	*	*	*	*		*
Turbidity						*	*	*	*	
Zinc			*							
Cryptosporidium							*			
Coliforms		*	*	*			*	*	*	*
Lead	*	*	*	*	*	*	*	*	*	*
Mains										
Iron		*	*	*	*	*	*	*	*	
Manganese			*			*		*		
pH								*		
Turbidity		*					*	*		
Aluminium		*		*			*	*		
Colour			*			*		*		
investigate	*									*
remedy	*	*								*

Table IV: Parameters for which Undertakings have been accepted from Water Service Companies
[? = insufficient data]

Iron relaxations have been granted widely, for example affecting supplies in over 30% of zones in areas of East Anglia and Yorkshire. Iron can occur in groundwaters derived from iron-rich strata such as Greensand and in waters from upland catchments, giving rise to turbid water and causing deposits in mains and customer complaints of discoloured water if adequate removal is not achieved in water treatment. Manganese is often associated with iron in these waters and relaxations for it are also widespread. Waters containing naturally-occurring aluminium, usually in low alkalinity, soft upland waters, sometimes combined with high colour, in many cases have complied with the earlier interpretation of the EC directive. Although some relaxations have been granted, many works are to be improved to achieve the more stringent interpretation in the Regulations by 1995 or earlier.

8 SECTION 20 UNDERTAKINGS

The Government's Medical Advisers, having considered the concentrations of the parameters subject to Section 20 Undertakings, have concluded that the Undertakings accepted in autumn 1989 pose no threat to public health. All the Water Services companies have Section 20 Undertakings for some parameters (Table IV). The significant ones are discussed below.

8.1 Distribution Systems.

All the companies have given undertakings to investigate or carry out work on distribution systems to overcome problems with some or all of the following parameters: iron, manganese, aluminium, colour, pH, turbidity or PAH. The Regulations have had a major impact in this area. In the past the "averaging" of sample results could enable the standards to be met, but this is no longer permitted.

Iron.
Iron can arise from corrosion of the main itself, but inadequate treatment of water at the works can also give rise to colour, iron, aluminium and manganese in distribution, leading to deposits and discoloured water. The undertakings have identified relining or replacement of mains in many areas. All the Water Services companies have programmes for renovation of their distribution systems by relining or replacement. Severn Trent Water estimate that over a 20 year period they may need to rehabilitate 70% of their distribution mains whilst for Anglian Water the figure is 38%. Other companies' estimates are between 4 and 13%. The costs of this work over the next 10 years are estimated by the Companies at £6 billion. There was serious doubt about the capacity of the pipeline renovation companies to undertake this amount of work and like several water suppliers, Wessex Water has set up its own commercial subsidiary company to develop these services.

Polynuclear aromatic hydrocarbons (PAH).
These compounds occur naturally in some untreated surface waters but can arise in excess of the limit of 0.2 ug/l in treated water from coal-tar bitumen linings of water mains. A Department of the Environment survey in early 1989 showed there was no widespread problem but further more-localised studies suggested a problem with particles of bitumen accumulating near deadends of distribution systems. Six of the plcs have given undertakings for PAH and may need to carry out replacement of pipe linings.

Lead.
The Water Quality Regulations require that where there is a significant risk of water failing to meet the lead standard due to lead from communication or service pipes, the water supplier must investigated how widespread the failures are in any zone. The test of a significant risk is whether more than 2% of samples taken at random from customers' taps fail to meet the standard in any zone (DoE 1989), adopting the sampling procedure for lead described above. These requirements were more stringent than the previous interpretation and water suppliers have not generally sampled in the manner now required. As a result, all the companies have given undertakings for lead. These affect all zones in North West, Southern, Welsh and Yorkshire Water (although this does not indicate that supplies throughout these areas are failing to meet the standard.) Anglian Water has undertaken to investigate the extent of non-compliance in 14% of its zones and to monitor the effect of existing treatment with orthophosphoric acid in another 66% of them. Where significant non-compliance is found water suppliers must investigate the effectiveness of water treatment in reducing plumbosolvency and install it if practicable, no later than 1995. The householders and public health authorities must be informed if water treatment is ineffective and lead pipe replacement is necessary.

8.2 Deficiencies at Service Reservoirs.

Undertakings for total coliform bacteria have been given for service reservoirs which fail to achieve 100% compliance with the zero standard. The actions to be taken include manual chlorination of water, installation of continuous chlorination equipment to maintain satisfactory residual concentrations and the prevention of ingress of rainwater through ducts, hatch covers or cracked walls or roofs. In some cases poor circulation of water in the reservoir is to be improved. In each case timescales are specified for completing these remedies, generally by 1992.

8.3 Deficiencies at Water Treatment Works.

Iron, manganese, aluminium and trihalomethanes.
Undertakings have been given for water supply zones which could be affected by parameters which are not adequately controlled in treatment. Aluminium, iron and manganese undertakings are common, affecting eight of the ten companies. One company has over 40% of its zones affected for iron and aluminium with problems at 24 out of 59 of its works. Remedial measures include continuous monitoring linked to autoshutdown in the event of concentrations over the limit, improvements to the design, operation and control of water treatment plant and the use of alternatives to aluminium coagulants. Failure to comply with the standard for trihalomethane, formed in water by the reaction of chlorine with organic matter, affects seven of the ten companies. Modifications to disinfection practices, alternative disinfectants and removal of precursors may be required. For frequent breaches of these standards the undertakings generally require compliance by the end of 1992; for occasional breaches the time-limit is the end of 1994.

Nitrate.
Excess nitrate in water supplies was considered by the U.K. Government to be "man-made" rather than due to the nature of the ground and hence it has been the subject of undertakings rather than relaxations. The companies affected are Anglian, where 28% of the sampling zones were covered, Severn Trent (5%), Yorkshire (1%) and Thames Water, where 11% of the treatment works required undertakings. Remedial measures include blending with low-nitrate water, treatment to remove nitrate and prevention of nitrate entering supplies by catchment protection measures. Most improvements are to be by 1992, with the remainder between 1993 and 1995.

Nitrite.
Nitrite in distributed water usually arises from microbiological action on ammonia which may be added to form chloramine - a more stable form of disinfectant. Levels of nitrite above the standard have been experience in distribution systems and undertakings exist for Anglian (53% of zones), Thames (21%), Welsh (42%), Yorkshire (2%) and North West. Improvements are to be completed by 1995.

Pesticide.
Pesticides (including herbicides and fungicides) have been found in very low concentrations in water over a wide area. Before July 1989 the U.K. Government's advice to water suppliers was to monitor for those compounds which were likely to be used in the catchment in addition to some environmentally persistent products banned from use some years ago. The EC standard of 0.1 ug/l for any single compound was not related to the varying toxicities of the wide range of products in use. The Government advice therefore was that provided concentrations did not exceed guideline values based on the toxicity, there was no risk to health. The health advice has not altered, but with the Regulations came the requirement to comply with the EC standard. Undertakings have therefore been given for pesticide by six of the ten companies, affecting from 1% to 60% of the sampling zones. Two companies have insufficient data to judge the extent of compliance. The terms of the undertaking are

to immediately inform the National Rivers Authority to enable it to investigate the possibility of controlling the leaching of pesticides into water;

for the company to immediately investigate suitable water treatment methods to remove the pesticides;

in the light of these investigations to install water treatment plant as soon as practicable, but generally by 1999.

The lack of information on suitable pesticide treatment processes and the absence of a relationship between the EC standard and toxicity of the compounds has led several EC member states to conclude that the limit is not currently achievable. The U.K. Government has requested the European Commission to review the standard.

9 COST OF IMPROVEMENTS

The cost of known requirements for improvements was estimated by companies in conjunction with Government-appointed auditors (Table V). Where there were known problems but their extent or solutions were unknown in autumn 1989, the Government agreed that these costs could not be included but might be passed on to customers via regulated charges at a later date. The £10 billion programme does not include the cost of remedial measures for much of the PAH, pesticide or lead problems.

Water Supply Function	£million
Water resources	985
Water treatment	3,060
Distribution systems	5,985
Total	10,030

Table V: Summary of water supply investment 1990 - 2000.
(November 1989 prices).

10 CONCLUSIONS

The quality and monitoring of water supplies in England and Wales is now specified by water quality regulations, which incorporate the standards of the EC Drinking Water directive and some additional standards, and specify the frequency of sampling and analysis. Supplies are required to meet the majority of these standards for every sample examined. Compared with an earlier interpretation of the Directive, the stricter standards have resulted in some supplies failing to comply. The Water Services companies established by the privatisation of the ten regional water authorities have been granted relaxations of some parameters for defined areas and for others the Secretary of State has accepted undertakings from them to carry out improvements to an agreed timetable. The most widespread non-compliance is for iron and related parameters from distribution systems and aluminium, iron and pesticide at treatment works. Investigation of potential non-compliance with the more stringent lead standard is required due to lead from domestic plumbing materials. Programmes of improvement should be completed for most water treatment problems by 1995 but the scale of the programme of water mains rehabilitation will require at least ten years for its completion. The estimated cost for problems with known solutions is £10billion for England and Wales.

11 REFERENCES

DoE 1982: Department of the Environment/Welsh Office. Circular 20/82: EC Directive relating to the Quality of Water intended for Human Consumption (80/778/EEC). Her Majesty's Stationery Office, London.

DoE 1989: Guidance on Safeguarding the Quality of Public Water Supplies. Her Majesty's Stationery Office, London.

EC 1975: Council Directive 75/440/EEC. Quality required of surface water intended for the abstraction of drinking water. Official Journal No. L194, 25.7.1975, p34.

EC 1980: Council Directive 80/778/EEC. Relating to the quality of water intended for human consumption. Official Journal No.

HMSO 1989a: Water Act 1989. Chapter 15. Her Majesty's Stationery Office, London.

HMSO 1989b: The Water Supply (Water Quality) Regulations 1989, Statutory Instrument 1989 No.1147. Her Majesty's Stationery Office, London.

HMSO 1989c: The Water Supply (Water Quality) (Amendment) Regulations 1989 Statutory Instrument No.1384. Her Majesty's stationery Office, London.

Miller 1990: Paper from Eureau 1 meeting, Paris, January 1990. Adaptation of Drinking Water Directive. Comments on Individual Parameters. D.G.Miller, Water Reserch Centre.

Schroder 1989: The Water Share Offer, Pathfinder Prospectus. J. Henry Schroder Wagg & Co. London.

W.H.O. 1984: Guidelines for Drinking-water Quality, Vol.1 Recommendations. World Health Organisation, Geneva.

ACKNOWLEDGEMENT

The author acknowledges the contribution made to this paper by many unnamed employees of the former water authorities and their advisers in drafting the Prospectus for the Water Share offer. The presentation of this paper has been approved by Gareth Jones, Director of Science and Quality, Wessex Water Business Services Ltd., but the views expressed are the author's and do not necessarily represent those of the Company.

APPENDIX

SCHEDULE OF WATER SUPPLY QUALITY STANDARDS FOR ENGLAND AND WALES

Average trihalomethanes (the sum of trichloromethane, dichlorobromomethane, dibromochloromethane and tribromomethane) not to exceed 100 ug/l; if less than four samples per year each sample not to exceed the limit.

PRESCRIBED CONCENTRATIONS OR VALUES

TABLE A

Item	Parameters	Units of Measurement	Concentration or Value (maximum unless otherwise stated)
1.	Colour	mg/l Pt/Co scale	20
2.	Turbidity (including suspended solids)	Formazin turbidity units	4
3.	Odour (including hydrogen sulphide)	Dilution number	3 at 25°C
4.	Taste	Dilution number	3 at 25°C
5.	Temperature	°C	25
6.	Hydrogen ion	pH value	9.5 5.5 (minimum)
7.	Sulphate	mg SO$_4$/l	250
8.	Magnesium	mg Mg/l	50
9.	Sodium	mg Na/l	150(i)
10.	Potassium	mg K/l	12
11.	Dry residues	mg/l	1500 (after drying at 180°C)
12.	Nitrate	mg NO$_3$/l	50
13.	Nitrite	mg NO$_2$/l	0.1
14.	Ammonium (ammonia and ammonium ions)	mg NH$_4$/l	0.5
15.	Kjeldahl nitrogen	mg N/l	1
16.	Oxidizability (permanganate value)	mg O$_2$/l	5
17.	Total organic carbon	mg C/l	No significant increase over that normally observed
18.	Dissolved or emulsified hydrocarbons (after extraction with petroleum ether); mineral oils	µg/l	10
19.	Phenols	µg C$_6$H$_5$OH/l	0.5
20.	Surfactants	µg/l (as lauryl sulphate)	200
21.	Aluminium	µg Al/l	200
22.	Iron	µg Fe/l	200
23.	Manganese	µg Mn/l	50
24.	Copper	µg Cu/l	3000
25.	Zinc	µg Zn/l	5000
26.	Phosphorus	µg P/l	2200
27.	Fluoride	µg F/l	1500
28.	Silver	µg Ag/l	10(ii)

Note (i) See regulation 3(5).
      (ii) If silver is used in a water treatment process, 80 may be substituted for 10.

TABLE B

Item	Parameters	Units of Measurement	Maximum Concentration
1.	Arsenic	µg As/l	50
2.	Cadmium	µg Cd/l	5
3.	Cyanide	µg CN/l	50
4.	Chromium	µg Cr/l	50
5.	Mercury	µg Hg/l	1
6.	Nickel	µg Ni/l	50
7.	Lead	µg Pb/l	50
8.	Antimony	µg Sb/l	10
9.	Selenium	µg Se/l	10
10.	Pesticides and related products:		
	(a) individual substances	µg/l	0.1
	(b) total substances(i)	µg/l	0.5
11.	Polycyclic aromatic hydrocarbons(ii)	µg/l	0.2

Notes (i) The sum of the detected concentrations of individual substances.

(ii) The sum of the detected concentrations of fluoranthene, benzo 3.4 fluoranthene, benzo 11.12 fluoranthene, benzo 3.4 pyrene, benzo 1.12 perylene and indeno (1,2,3-cd) pyrene.

TABLE C

Item	Parameters	Units of Measurement	Maximum Concentration
1.	Total coliforms	number/100 ml	0(i)
2.	Faecal coliforms	number/100 ml	0
3.	Faecal streptococci	number/100 ml	0
4.	Sulphite-reducing clostridia	number/20 ml	≤ 1(ii)
5.	Colony counts	number/1 ml at 22°C or 37°C	No significant increase over that normally observed

Notes (i) See regulation 3(6).

(ii) Analysis by multiple tube method.

TABLE D(i)

Item	Parameters	Units of Measurement	Maximum Concentration or Value
1.	Conductivity	µS/cm	1500 at 20°C
2.	Chloride	mg Cl/l	400
3.	Calcium	mg Ca/l	250
4.	Substances extractable in chloroform	mg/l dry residue	1
5.	Boron	µg B/l	2000
6.	Barium	µg Ba/l	1000
7.	Benzo 3,4 pyrene	ng/l	10
8.	Tetrachloromethane	µg/l	3
9.	Trichloroethene	µg/l	30
10.	Tetrachloroethene	µg/l	10

Note: (i) See regulation 3(3)(d).

TABLE E

Item	Parameters	Units of Measurement	Minimum Concentration(i)
1.	Total hardness	mg Ca/l	60
2.	Alkalinity	mg HCO₃/l	30

Note: (i) See regulation 3(2).

Chapter 5

CONTROL OF DANGEROUS SUBSTANCES IN UK SURFACE WATERS

T F Zabel (Water Research Centre, UK)

ABSTRACT

The Environmental Quality Objective (EQO) approach is being widely applied in the UK for the control of discharges to surface waters. The basis of the EQO approach is the definition of beneficial uses of fresh and saline waters and the setting of Environmental Quality Standards (EQSs) to protect the uses with consent conditions for discharges being set so as to meet the relevant EQSs in the receiving waters. However, in response to the Ministerial Conference on the Protection of the North Sea in 1987, the UK has adopted the dual approach (applying uniform emission standards (UESs) and EQSs whichever are the more stringent) for the control of particularly dangerous "Red List" substances. The paper describes the approach adopted in the UK for the control of dangerous substances, with special emphasis on the control of the "Red List" substances, and the procedure used in the UK for deriving EQS values and for setting limit values for effluent discharges to meet the EQSs in the receiving waters.

1. INTRODUCTION

Although legislation to control pollution of surface water dates back to the Public Health Act of 1875, which made it an offence to pollute certain waters, the EC Directive on the Control of Dangerous Substances (76/464/EEC) has been the most far-reaching measure to achieve a reduction or elimination of pollution of surface waters by dangerous substances (CEC 1976). This paper summarises the legislation adopted in the UK to control pollution and discusses the approaches used in the UK to control the discharge of dangerous substances to the aquatic environment.

2. UK POLLUTION CONTROL LEGISLATION

In the UK, Parliament adopts the enabling legislation to control pollution including discharges to sewers. The legislation provides the general framework for controls

Water Treatment – Proceedings of the 1st International Conference, pp. 43–54

and gives the Secretary of State for the Environment the power to issue orders giving instructions for the implementation of the legislation. The orders are issued as commencement orders or 'statutory instruments'. The procedure for changing laws in the UK is to repeal certain sections of existing legislation and to issue new legislation in Acts of Parliament. Thus some sections of the Public Health Act 1875 which controlled discharges to surface waters are still in force whereas most sections have been superseded by new Acts, the most important of which are the Control of Pollution Act 1974 and the Water Act 1989. Similarly, although the control of discharges to sewers refers back to the Public Health Act of 1936, most of the provisions in the Act have been superseded by the Control of Pollution Act 1974 and the Water Act 1989.

The Control of Pollution Act 1974 strengthened the provisions for the control of pollution, in particular the control of discharges to estuarine and coastal waters. The consent conditions for discharges to surface waters and the monitoring data were made available for public scrutiny. The data can be used by the public to initiate prosecutions. Applications for new consents need to be advertised allowing objections to be raised to the consent conditions proposed.

The Water Act 1989 set up the National Rivers Authority in England and Wales which now has the responsibility for the monitoring of discharges to surface waters, both those from the newly privatised water companies and industry (previously the treatment of wastewaters and the monitoring functions were carried out in England and Wales by the same organisation, the Water Authorities). The Water Act 1989 also requires the National Rivers Authority to define statutory quality objectives for all surface waters. In Scotland the treatment and monitoring functions have been separate since 1951 with the Regional Councils being responsible for treatment and the River Purification Boards for the monitoring.

The use-related EC directives, eg for the abstraction for drinking water (75/440/EEC) (CEC 1975), were implemented in the UK by administrative action which involved the Department of the Environment (and/or the Welsh Office and the Scottish Development Department) writing to the responsible bodies (at the time Water Authorities in England and Wales and River Purification Boards in Scotland) advising them on the contents of the directives and the actions necessary to comply with them. In Northern Ireland, the Department of Environment for Northern Ireland is itself the responsible body.

As the Dangerous Substances Directive (CEC 1976) is essentially an enabling directive, no immediate requirements to comply were needed. However, following the publication of several 'daughter' directives for individual substances and the setting of environmental quality standards (EQSs) in the UK for several List II substances, the Department of the Environment and the Welsh Office published in 1985 a circular (DoE 1985) which advised the responsible authorities at the time (the Water Authorities in England and Wales) of the content of the Dangerous Substances Directive and the 'daughter' directives then published. The circular also implemented the standards laid down in the directives and those agreed nationally for the List II substances (arsenic, chromium, copper, inorganic lead, nickel and zinc). A similar circular was issued by the Scottish Development Department to the River Purification Boards in Scotland (SDD 1985). This DoE circular has recently been superseded by the updated Circular 7/89 (DoE 1989) implementing all directives adopted so far for List I substances and EQSs for an additional six List II substances or groups of substances (vanadium, boron, organotins, mothproofing agents, pH and iron).

3.   THE ENVIRONMENTAL QUALITY OBJECTIVE (EQO) APPROACH

The EC Dangerous Substances Directive (CEC 1976) provides two alternative approaches for control, the uniform emission standard (UES) or limit value approach and the environmental quality standard (EQO) approach.

The EQO approach has been generally applied in the UK for the management of discharges to surface waters. The EQO approach is based on the premise that a minimum acceptable concentration of a pollutant can be defined which does not interfere with the use of the water. This premise does not hold for certain particularly dangerous compounds, which require different control strategies. For instance, DDT control demands prohibition of use or total ban on discharges (complete recycling or destruction).

A distinction has to be made between EQOs and EQSs. The EQO defines the use for which the water is intended (eg abstraction for drinking water) whereas the EQS specifies the concentration of the substance which should not be exceeded to protect the particular use of the water. For example, in the Abstraction for Drinking Water Directive the EQS for nitrate is 50 mg/l.

In the EC Dangerous Substances Directive environmental quality standards (EQSs) are described as environmental quality objectives (EQOs). This has led to confusion as quality objectives are viewed by those countries which opted for the limit value approach as long-term goals demanding very low standards and not as in the UK for setting consents for discharges.

To apply the UK EQO approach, the different uses of water need to be defined and standards, in terms of the maximum acceptable concentration of the particular contaminant, must be derived. These concentrations must be low enough to protect the water's uses by taking into account the vulnerable targets requiring protection: man, his food sources or other living organisms. The standards can then be used to derive consents for individual discharges, taking into account the dilution in the receiving water, but also the concentration of the substance already present in the receiving water from other sources. The EQO approach can deal with diffuse inputs and can be used to control multiple point sources. It also makes proper use of the receiving water capacity. The common uses of both fresh and saline waters have been defined (Mance 1984), taking into account the use-related directives adopted by the EC (eg the Abstraction to Drinking Water Directive (CEC 1975) and the Freshwater Fish Directive (CEC 1978), Table 1.

Table 1.   Designated uses of surface waters for the derivation of EQS values

Use	Freshwater	Saline water
For direct abstraction to potable supply	/	-
For abstraction into impoundment prior to potable supply	/	-
As a source of food for human consumption	/	/
Protection of fish and shellfish	/	/
Protection of other aquatic life and dependent non-aquatic organisms	/	/
Irrigation of crops	/	-
Watering of livestock	/	-
Industrial abstraction	/	/
Bathing and water-contact sports	/	/
Aesthetic considerations	/	/

- Not applicable

In practice, however, the setting of a reliable EQS for all water uses is frequently not possible as the information on harmful effects is often limited.

4.   DERIVATION OF ENVIRONMENTAL QUALITY STANDARDS (EQSs)

Figure 1 illustrates the approach adopted for the derivation of EQSs in the UK.

Available data on the toxic effects, both from laboratory studies and field observations, are critically assessed. Particular attention is paid to the experimental test method used.

The lowest credible adverse effect concentration is derived from the available toxicity data. Particular emphasis is placed on chronic effects noted after long-term exposure, or exposure effects at specific sensitive life stages of the target species. A safety factor is then applied to obtain a tentative EQS, which depends on the quality and extent of the available toxicity data. If only acute (short term) LC_{50} values are available, an arbitrary safety factor of 100 is usually applied. This factor tends to be reduced to 10 if chronic LC_{50} data are available for sensitive species. For some substances, 'no effects concentrations' (NOEC) are reported for long-term chronic studies. These values may be used as tentative standards without applying a safety factor, provided the test design is acceptable, and that they refer to relevant species and are supported by the other data available.

The tentative EQS derived from the laboratory toxicity data is verified by comparison with the available field toxicity data, taking care to distinguish between the effects of the particular compound and those of other potentially toxic compounds

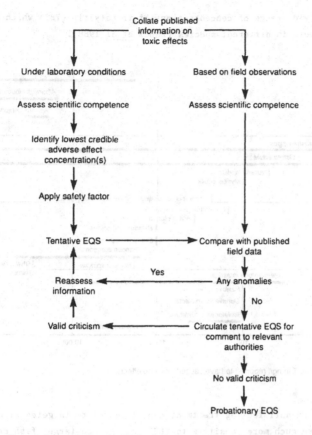

Figure 1. Schematic flow chart for the derivation of environmental quality standards.

present. It is important to compare the tentative EQS with the highest observed concentration which has no effect on the ecosystem, rather than with the lowest concentration which does have an effect. Any discrepancies between the field data and the tentative EQS require the reassessment of the available field and laboratory data and, if necessary, the application of a different safety factor.

The setting of EQS values relies heavily on knowledge of bioaccumulation and toxicity effects. This information is often incomplete and depends on interpretation. The field data are often inadequate, especially for the less frequently occurring organic compounds, to verify the tentative EQS derived from the laboratory data.
The application of safety factors must therefore take account of the different toxicities and actions of the chemicals on the target organisms and the quality of the toxicity data available.

When deriving EQS values for substances which are designed for a specific purpose, such as pesticides, it is particularly important that toxicity tests are available for the target species as the sensitivities of different taxonomic groups to these compounds vary widely.

Figure 2 shows the wide range of concentrations of tributyltin (TBT) which have caused adverse effects in different species (Zabel et al 1988).

Fish
- Solea solea
- Alburnus alburnus
- Agonus cataphractus

Molluscs
- Crassostrea gigas
- Ostrea edulis
- Nucella lapillus
- Mytilus edulis
- Venerupis decussata
- Venerupis semidecussata

Crustacea
- Gammarus oceanicus
- Acartia tonsa
- Balanus balanoides
- Elminius modestus
- Orangon crangon
- Nitrocra spinipes
- Rhithropanopeus harrisii

Algae
- Enteromorpha intestinalis
- Pavlova lutheri
- Dunaliella tertiolecta
- Skeletonema costatum
- Thalassiosira pseudonana

| 1 | 10 | 100 | 1000 | 10 000 |

Figure 2. Concentrations of TBT in ng/l reported to have caused adverse effects.

TBT is designed as an anti-fouling treatment and is therefore targeted at molluscs and algae, which are much more sensitive to TBT than the non-target fish species. However, there is also a wide range of sensitivities for different species of the same taxonomic group, and it is therefore important to collect data for more than one species.

Bioaccumulation is another important parameter for assessing 'safe concentrations' in the environment since some compounds can be selectively retained in the living tissue of animals and plants. These compounds can cause direct effects on the target organism, or they may be transferred via the food chain to other organisms. They can also lead to discoloration or tainting of the tissues, making them unacceptable for human consumption or causing their rejection as food by higher organisms.

Bioaccumulation was utilised in the derivation of the marine environment EQS for TBT. The correlation between the incidence of imposex (caused by growth blocking the pallial oviduct of females, preventing the release of eggs and causing premature death) in the dog whelk, Nucella lapillus, and the corresponding TBT tissue concentrations was determined. The bioaccumulation factor was used to determine the water concentration which would produce an insignificant increase in the occurrence of imposex. This water concentration was similar to the tentative EQS derived by applying an arbitrary safety factor of 10 to laboratory toxicity test results. It was thus possible, using bioaccumulation, to verify the tentative EQS.

When setting EQSs, it is important to establish whether the standard should be set as total or dissolved concentration or, as in the case of ammonia, the un-ionised species. This approach requires a knowledge of the speciation of the compound in the environment, and is of particular importance for inorganic compounds. For organic compounds, the EQS tends to be expressed as total concentration.

A decision has also to be made on the degree of compliance required for the standard to protect the use. The options include:

- 100 per cent compliance (maximum allowable concentrations)
- 95 per cent compliance (5 per cent exceedance can be tolerated)
- 50 per cent compliance (annual mean concentration)

The decision depends on the source of the substance and on the difference between acute and chronic toxicities.

Table 2 shows the List II substances for which EQS values have been adopted in the UK and another 11 compounds which are currently under discussion.

The rationale for the different standards for the individual substances is provided in the appropriate Water Research Centre Technical Reports. The same methodology is also being applied to derive Preliminary Environmental Quality Standards (PEQSs) for

Table 2. Substances for which EQS values have been published

Substances	WRc report number TR
Chromium	207
Inorganic lead	208
Zinc	209
Copper	210
Nickel	211
Arsenic	212
Vanadium	253
Organotins	255
Boron	256
Iron	258
pH	259
Mothproofing agents	261
Substances currently under discussion	
Inorganic tin	254
Sulphide	257
Ammonia	260
Trichlorobenzenes	
Dichlorobenzenes	
Monochlorobenzene	
Xylenes	
Toluene	
Benzene	
Atrazine	
Simazine	

those 'Red List' substances (see Section 5) for which EQS values have not yet been adopted. The standards adopted are periodically reviewed based on new information published in the literature. Currently the standards set in 1984 for the six List II metals (arsenic, chromium, copper, lead, nickel and zinc) are being reappraised.

5.   CONTROL OF PARTICULARLY DANGEROUS 'RED LIST' SUBSTANCES

By insisting on applying the EQS approach to the control of dangerous substances, whereas the other EC Member States opted for the uniform limit value (UES) approach, the UK has been accused of being lax on the control of environmental pollution. This is partly based on the misconception that the EQS is being used to allow discharges to utilise the full capacity of the receiving water. In reality the EQS is used by the controlling authority for estimating the upper limit for a consent and, more importantly, as the indicator of acceptable water quality for a specified use - a factor which is not included in the UES approach. In practice the controlling authority is likely to restrict the discharge for a new site to less than 20% of the residual available capacity to minimise the potential impact on the receiving water.

In its report on the EC 'framework' Directive, the House of Lords Select Committee on the European Communities suggested that Member States should re-examine the alternative use of limit values and quality objectives (House of Lords 1985). This proposal was endorsed by the House of Commons Select Committee on the Environment (House of Commons 1987).

In the Ministerial Declaration following the Second Ministerial Conference on the Protection of the North Sea held in November 1987, the UK agreed to apply the dual approach to the control of certain particular dangerous substances (DoE 1987). The proposal recommends:

- identification of a limited range of the most dangerous substances based on scientific criteria - the 'Red List'
- setting of strict environmental quality standards for all 'Red List' substances
- progressive application of emission standards based on the concept of 'best available technology not entailing excessive costs' (BATNEEC) for the control of direct discharges of 'Red List' substances.

The aim of the new proposals is that an effluent containing 'Red List' substances will have to meet industry-wide effluent standards as well as the EQS for the receiving water to ensure that the effluent discharged does not cause adverse effects. Best available technology will initially be applied to new and refurbished plants handling any of the 23 'Red List' substances given in Table 3 (DoE 1989).

In accordance with the Ministerial Declaration the UK has also undertaken to reduce inputs of the 'Red List' substances to UK coastal waters by about 50%. This might be achieved by applying strict EQS and limit values, but could also involve additional control measures such as reduction in use, particularly for diffuse inputs. The UK

Table 3. UK 'Red List' substances

	EC Directive adopted (List I status)
Mercury	+
Cadmium	+
*gamma - hexachlorocyclohexane (lindane)	+
*DDT	+
*Pentachlorophenol (PCP)	+
Hexachlorobenzene (HCB)	+
Hexachlorobutadiene (HCBD)	+
*Aldrin	+
*Dieldrin	+
*Endrin	+
*PCB (Polychlorinated biphenyls)	
*Tributyltin compounds	
*Triphenyltin compounds	
*Dichlorvos	
*Tricluralin	
1,2 Dichloroethane	
Trichlorobenzene	
*Azinphos-methyl	
*Fenitrithion	
*Malathion	
*Endosulfan	
*Atrazine	
*Simazine	

* Substances which enter the environment predominantly by indirect routes

has initiated a monitoring programme to determine the baseline load discharged to provide proof for the success of the reduction programme. Work is also in progress to identify the major sources of 'Red List' substances. The information will be used to develop action programmes to achieve the desired reduction of approximately 50% by 1995.

Her Majesty's Inspectorate of Pollution (HMIP) is responsible for developing 'BATNEEC' for specified industries based on the integrated pollution control (IPC) concept. The aim of this approach is to minimise pollution of all environmental compartments by making the same authority responsible for controlling discharges to all three environmental compartments (air, water, soil). Consents for discharges to water will be derived in consultation with the National Rivers Authority to ensure that the EQS values in the receiving water are not exceeded.

6. CONCLUSIONS

- The control of dangerous substances in the UK is based on the EQO/EQS approach which involves setting appropriate standards to protect the different uses of water.
- The EQS values are used to derive consents for discharges taking into account the concentration of the substance already present in the receiving water from diffuse and other point sources.

- For particularly dangerous substances, the 'Red List', the UK has adopted the dual approach of control using limit values based on BATNEEC and EQSs whichever are the more stringent.
- For the control of 'Red List' substances, the integrated pollution control (IPC) approach is being applied in an attempt to deal with the cross-sectoral aspects of waste disposal. The effects on all three environmental compartments are being assessed when issuing a consent for a particular process.
- The aim of the quality objective approach is to move progressively towards higher goals, including the minimisation of inputs of the most dangerous substances to the aquatic environment, taking advantage of technical advances where practical.
- The quality standards are viewed as the minimum to be achieved and the aim is for a quality as well within that standard as possible. The standards must not be used to relax the control of discharges simply because the standard is easily met.

REFERENCES

COUNCIL OF EUROPEAN COMMUNITIES (1975) Council directive concerning the quality required of surface water intended for the abstraction of drinking water in Member States (75/440/EEC), Official Journal L194, 25 July.

COUNCIL OF EUROPEAN COMMUNITIES (1976) Council Directive on pollution caused by certain dangerous substances discharged into the aquatic environment of the Community (76/464/EEC), Official Journal L129, 18 May.

COUNCIL OF EUROPEAN COMMUNITIES (1978) Council Directive on the quality of freshwaters needing protection or improvement in order to support fish life (78/659/EEC), Official Journal L222, 14 August.

DEPARTMENT OF THE ENVIRONMENT AND WELSH OFFICE (1985) Water and the environment. Circular 18/85 (Circular 37/85 Welsh Office), 2 September.

DEPARTMENT OF THE ENVIRONMENT (1987) Ministerial Declaration, 'Second international conference on the protection of the North Sea', London, 24-5 November 1987.

DEPARTMENT OF THE ENVIRONMENT AND WELSH OFFICE (1989) Water and the environment. Circular 7/89 (Circular 16/89 Welsh Office) 30 March.

DEPARTMENT OF THE ENVIRONMENT (1989) Agreed 'Red List' of dangerous substances confirmed by the Minister of State (Lord Caithness). News Release 194, DoE, Marsham Street, London, 10 April.

HOUSE OF COMMONS SELECT COMMITTEE ON THE ENVIRONMENT (1987) House of Commons Session 1986/87. 3rd Report from the Environment Committee (HC 183). HMSO, London.

HOUSE OF LORDS SELECT COMMITTEE ON THE EUROPEAN COMMUNITIES (1985)   Dangerous substances - Session 1984-85, 15th Report.   HMSO, London, 23 July.

MANCE G (1984)   Derivation of environmental quality objectives and standards for black and grey list substances.   Chem Ind (London), 509.

SCOTTISH DEVELOPMENT DEPARTMENT (1985)   Implementation of EC Directive 76/464/EEC: Pollution caused by certain dangerous substances discharged into the aquatic environment of the Community.   Circular 34/1985, 29 November.

ZABEL T F, SEAGER J and OAKLEY S D (1988)   Proposed environmental quality standards for List II substances in water: organotins.   Water Research Centre Technical Report TR 255, Medmenham.   March.

PART II

Design and construction of plant

Chapter 6

THE GREATER LYON EMERGENCY WATER INSTALLATIONS

M Delaye and C Abgrall (Compagnie Générale des Eaux, Lyon, France)

THE VULNERABILITY OF THE LYONS CONURBATION

Greater Lyons is supplied with practically the whole of its drinking water from the underground river situated beneath the River Rhône. A huge area of 300 hectares, located North-East of Lyons and 5 kilometres from the Place Bellecour, contains over 130 wells in operation ; water infiltrates from the river and alluvial deposits, which possess excellent filtering properties, ensure that this water is purified by a natural process.

Correct operation of this water catchment system is thus highly dependent on the quality of water from the River Rhône, and could be endangered by any accidental dumping of chemicals in the river. It is common knowledge that the risk of this type of accident cannot be completely excluded.

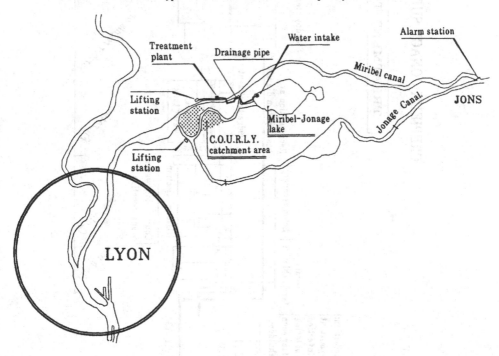

Water Treatment – Proceedings of the 1st International Conference, pp. 57–60

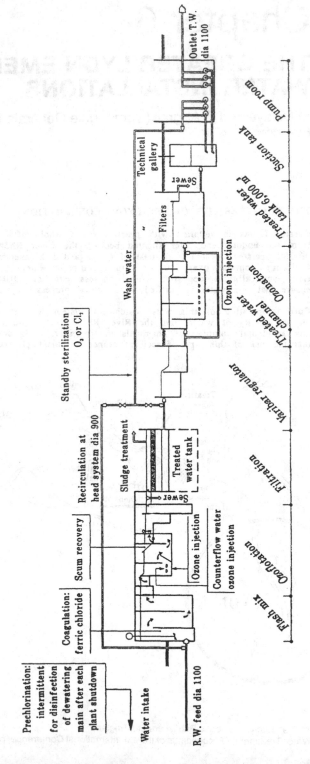

CRÉPIEUX JONAGE SUPPLY SYSTEM

TREATMENT TRAIN

THE DECISION TO CONSTRUCT EMERGENCY WATER INSTALLATIONS

For this reason, in 1984 the Greater Lyons Council decided to set up emergency resources, which would make it possible to interrupt pumping operations in the catchment area during the passage downriver of the pollution slick. The capacity of the emergency water installations (150,000 cubic metres/day) was defined following studies carried out on the risks of pollution and the corresponding durations of interruptions to water gathering operations, it being understood that low-priority water consumation (for the purposes of watering, road washing etc...) will be suspended during the passage downriver of the pollution.

Amongst the different solutions examined, that involving treatment of water from Lake Miribel-Jonage quickly emerged as the best choice, from both a technical and economic standpoint. This lake, recently created, is a result of careffully-coordinated extraction of alluvial deposits as part of a project for the development of a site for leisure activities. Located to the North-East of Lyons, it features a capacity of 7 000 000 cubic metres (which will eventually be increased to 12 000 000 cubic metres) of water with highly satisfactory bacteriological and chemical characteristics.

Given the geographical location of the lake, the logic behind the project is quite straightforward : Untreated water is drawn from the South-West side of the lake, and sent by a pumping-station along a pipe 1,10 metres in diameter to the treatment-plant, which is located in Rillieux, next to the Miribel canal. Treated water is returned to the catchment field pumping network, which means that the primary pumping installations of the conurbation can be used.

The development of this project was entrusted to the "Compagnie Générale des Eaux", as part of a concession agreement for construction and operation of the project.

TREATMENT PROCESS

The treatment process is of course adapted to suit characteristics of lake water.

The Rillieux plant is thus the first installation of its size to use the ozoflotation technique developed by OTV, which is particularly suited to eliminating algae and other suspended particles : although Lake Miribel does not for the time being give rise to any worries on this account, algae development remains a constant threat to lake water.

This process, which is carried out at the start of treatment operations, comes after flocculation with ferric chloride followed by flash mixing. The principle is as follows (see diagram below). Water overflow into the ozonation compartment : porous plates diffusing ozone are swept along by a stream of water, causing tiny bubbles to be created. This tiny bubbles are dragged into the second compartment (flotation compartment), trapping a very hight proportion of floating particles and ensuring optimum ozone transfer. The treated water exits from base of the flotation compartment and rejoins the outlet channel.

Periodic evacuation of impurities is carried out by raising the level of the water and using a transversal stream of water to sweep away the foam.

The water is subsequently filtred by a battery of 9 twin-layer "anthracite-sand" filters, after which it undergoes a final ozonation process before entering the two 3 000 cubic metre tanks for treated water. These tanks constitue the buffer storage of the installation and are used for bash washing the filters as well as for immediate start-up of the treatment plant.

Sludge coming from the ozoflotation process and from the filters is treated with the aid of a revolving blade thickener/clarifier followed by a belt-filter ; this process makes it possible to reach a dry content value of over 25 %.

One aspect related to the production of ozone is worthy of note : the ozone used here is produced not from air but from pure oxygen, via a battery of two ozonizing devices.

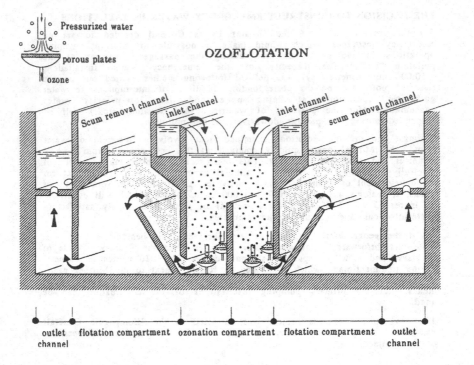

Pressurized water

porous plates

ozone

OZOFLOTATION

Scum removal channel   inlet channel   inlet channel   scum removal channel

outlet   flotation compartment   ozonation compartment   flotation compartment   outlet
channel   channel

This original solution, which makes for easier start-up of the plant, is particulary suited to an emergency installation destined for sporadic use.

OPERATION OF THE PLANT

To guarantee perfect reliability -a vital objective for emergency equipment- the installation will operate four times a week in normal conditions. In addition, a water recirculation system has also been developed, enabling the plant to operate using its own water ; this is to ensure that the processed water stored in the thanks remains at an optimal level of quality and to authorise the immediate start-up of supply by the plant.

The speed of operation of this type of emergency installation is obviously of the utmost importance. In the event of pollution, the Rillieux plant, which is completely automated, will be put into service by remote-control from the central Croix-Luizet Station, which also controls the other production installations and the pumping-stations.

Accidental pollution will be detected by a warning station equipped with continuous analysers, located in Jons (20 kilometres upstream of the catchmen area). The presence of this station however serves only to complement the usual channels for transmission of an alert which will signal the presence of pollution before it reaches Jons.

Collaboration with industrialists, which began at the moment of initial risk studies, is thus bound to continue. Such collaboration will also make it easier to handle crisis situations, thanks to an improved knowledge of possible polluants and the fact that the various participants have been made aware quickly and precisely.

In conclusion, the development of the Rillieux emergency plant undoubtedly constitues a major step forward regarding a reliable supply of drinking water for the Lyons conurbation. A technological first thanks to its ozoflotation process it represents a break - for reasons of safety - with an age-old tradition in Lyons, namely the exclusive use of underground water.

Chapter 7

EXPERT SYSTEMS FOR THE EXPLOITATION OF A PURIFICATION STATION: ORAGE (COMPUTER AID FOR THE MANAGEMENT OF PURIFICATION)

M Mansat (Greater Lyon Council, Lyon, France)

Purification Station of MEYZIEU belonging to
the LYON Metropolitan Authority

Since its creation in 1969, the Sanitation Department of the Lyon Metropolitan Authority (COURLY) maintains preferential deelings with OTV Company (Omnium of Treatment and Valorization), a branch of the General Company of Water (C.G.E).

This Company has indeed co-operated in the building of two important purification stations of Courly :

- ST FONS on the left bank of the Rhône river (700 000 Eq/inhab.)

- PIERRE-BENITE on the right bank of the Rhône (475 000 Eq/inhab.).

OTV is also running, on behalf of the Sanitation Department and together with BTA Company, the purification Station of ST FONS.

Moreover, in 1988, OTV had been choosen as the pilot of a group including CGEE for Electricity and MAIA-SONNIER for civil engineering and in charge of the building of a new purification Station in MEYZIEU.

The filling of this station (capacity of 35 000 Eq/inhab. with a possibility of extention to 50 000 Eq/inhab.) has been done by the end of 1989.

This plant, modernly conceived and highly automated, has :

. a classical pre-treatment,
. a primary treatment made up of 2 lamellar decanting machines,
. a biological treatment with 4 underwater bacterial filters made by OTV
  and called BIOCARBONE.

The mud treatment is done via a thickening machine and 2 strip-filters aiming at desiccating muds before their dispatching to the purification station in PIERRE-BENITE where they will be burnt.

Moreover, the Sanitation Department, aiming at a strict management, tries to find means of curbing expenses at best and in particular its development problems.

So it was a good opportunity for innovating in the field of purification stations development, by putting together all skills of both OTV and COURLY'S Sanitation Department, the first one regarding technology and purification engineering and the latter for management and running of the Sanitation network and Stations.

© 1991 Elsevier Science Publishers Ltd, England
Water Treatment – Proceedings of the 1st International Conference, pp. 61–66

That's why, it came from a common agreement to set up, on the purification Station in MEYZIEU, an expert system which helps in taking decisions.

The Sanitation Department scope of powers :

With an annual budget of about 500 MF and a manpower of 625 persons, the Sanitation Department gets all powers regarding Sanitation on COURLY'S area :

- surveys and plans,
- building of networks and purification stations
- development of the network and purification stations.

In that way, it has to administrate :

- 2 200 km of unitary collectors, including 700 km which can be inspected,
-     25 stations for the pickup of waste water,
-      9 purification stations representing a total capacity of about 1 500 000 Eq/inhab.

Why implementing an expert system ?

To help its management, the Sanitation Department developed a special system to :

- curb and reduce operating costs,
- widen its knowledge about existing equipments and improve the ability of dealing with rainy periods.

These tools are :

- a mathematical device (SERAIL), allowing the rekoning of theoretic flows in every point of the Sanitation network during rainy periods of a given intensity,

- a computer file mentioning all manufacturers and their present or possible rejections in the network,

- an urban data base (SUR : Urban Reference System),

- a teleprocessing system for purification and pickup stations to which are given all pluviometric and exploitation measures,

- an analytic accountancy for calculating cost prices and ratios.

However sophisticated and performing tools may be, their implementation and use lead to two findings :

1) The knowledge about the network and stations operation has improved a lot but there are few or no means of having a rational influence on them : Storm Spillways are not, or slighthy, automated and present manufactured purification stations are not equipped with machines fitted to modify their output and their impact on natural environnement.

2) Information, measures and signs devices deliver so much data that it is impossible for the operating man to collect them all at a precise moment and then, even if he got them, he wouldn't have enough intellectual capacities to do the treatment and synthesis.

That's why in the purification station of MEYZIEU - which is equipped with the necessary machines able to modify its output and impact on environnement - it was decided to set up an expert system that will give the operating man, thanks to its computation power and its swiftness in synthesizing information, necessary elements to make the managing decisions.

Summary description of the station in Meyzieu (See diagram).

I - PART OF THE EXPERT SYSTEM :

Using a classical screen and a keyboard, the operator simulates the evolution of the station state for a day, a week or more.

The operator must give the following information to the system :

- the beginning date of simulation,
- scheduled precipitation (date, hour, intensity)
- number of decanting machines, biocarbone filters and strip - filters in use.

At first, the system displays on the screen the following graphs, according to the time (hour by hour and day by day) :

- flow at the station entrance

- BOD, COD and MS concentration in primary water, decanted water and treated water

- Mass of mud in the thickening machine

- Dates and hours when to do the washing of each biocarbone filter

- Operating duration and frequency of mud dehydration (and number of machines to be used simultaneously or not).

From these graphs, the operator can do, in a second time, simulations which will allow a better management of the station.

For each simutation, the same graphs - as above mentioned - will be displayed on the screen.

Thus, the operator can on demand :

. Simulate a bypass of primary water or decanted water (the bypass opens from 0 to 100 %).

. Simulate a stop (for maintenance, saving or lack of manpower) of 1, 2, 3 or 4 biocarbone filters, a lamellar decanting machine, 1 or 2 strip filters.

. Simulate an additional feed of water or pollution in the station.

According to the results given by the system during simulations, the operator can be led to take the following decisions :

- to manage the available manpower according to circumstances : cleaning out or dehydration tasks on most favourable dates ;

- to manage the washing of filters or to carry out "mini-wasking" at most propitious dates and hours : electricity costs, sanitation output, etc ...

- to manage the maintenance of the constituent elements of the station : a decanting machine stop for maintenance, a strip - filter stop, etc ...

- to be able to know, at any time and according to the pluviometry, on one hand which flow must be treated biologically and on the other hand by primary processing to respect the rejection standards ;

- to stop or set in action some biocarbone filters, some decanting machines, to open or shut more or less bypasses in order to save in particular the electrical energy consumption and also to change the purification ouput of the station according to the state of the river ;

- to avoid the muds to stay too long in the thickening machine and therefore the emission of bad odors ;

- to carry out the muds dehydration when they can be conveyed quickly to the purification station in PIERRE-BENITE to be burnt and when the furnace capacity is enough to receive them without any problem.

II - TECHNICAL DATA :

- a PS 2 IBM computer Model 70 with a keyboard

- number of rules : 100

- software : OTV - Inference motor NEXPERT
                    (Neurone - Data)

Except the rules' base of a classical expert system, algorhythms were drawn up by OTV Company. These algorhythms link entries and exits of the different plant's constituent elements (decanting and thickening machines, biocarbone filters). For example, in a biological treatment : to a given entry (flow, MS, BOD, COD) and a given clogging of the filter, you get a corresponding exit, and it occurs a given additional clogging.

III - FIGURES AND INFORMATION GIVEN TO THE SYSTEM :

When the system was first set up, due to a lack of measures really done, only the following graphs were entered in the system memory :

- flow = f (t) according to the weekday and a given intensity rain

* It is presently possible because the sanitation station of Meyzieu receives water from only one basin, of small capacity and relatively homogeneous.

- BOD, COD, SM pollution = f (t) according to the weekday and the rain.

These graphs are made out from statistics realized thanks to real measures taken on the network leading to the station.

Later on, the attribute of an expert system aiming at the data base development, the station collectors will improve graphs by giving real figures on flow and pollution.

Here are the scheduled collectors :

- Decanted water temperature

- Decanted water COD

- Treated water COD

- Primary water COD

- Decanted water SM

- Treated water SM

- Primary water SM

- Concentration of drawn off muds

- Concentration of thickened muds

- Mud film in the thickening machine

- Flow of Jonage Canal (river where the reject is done)

- Flow measure of primary water

- Flow measure of water accepted in primary treatment

- Flow measure of water accepted in biological treatment.

FUTURE PROSPECTS :

. The expert system set up in Meyzieu works most certainly on figures based on dry and polluted period flows that can be considered as reliable because they are established from statistics and real measures.
However, the influence of pluviometry on these values can't be determined in advance. The operator can only give, manually to the system, short term and relatively uncertain weather forecasts.

Later on, the use of radar will allow future projections to be more strict and more reliable.

- Moreover, we can expect that the implementation of the first system in the purification station of Meyzieu will be followed by others in the 8 remaining plants of COURLY.

It will be a means to manage the whole sanitation network, to privilege or not some feeding basins rather than others, in order to reduce the running cost and have a better control over rejection in natural environment.

The increasing control over the management of purification stations and storm spillways and the continuous and immediate knowledge about the pollution released in the environment, should prompt official authorities (Shipping Department, Basin Agencies, etc ...) to modify the rejection standards dictated to operators in sanitation networks. It seems indeed rational to have an optimum output in purification when the river receiving the reject is at its lowest level and though purification performances can drop when the river is in spate.

Finally, for the conception, realization and adjusting of this expert system, the OTV Company and the Sanitation Department of COURLY have done great financial and intellectual investments.

It would be a pity to keep this expert system confidential and without evolution ; therefore, OTV and the Sanitation Department of COURLY have set up an Economical Interest Group (GIE) which goals are :

- to promote this kind of system in other purification stations either in
  France or abroad ;

- to help the operators in defining their needs, in conceiving and
  adjusting this system helpful in taking decisions.

PART III

Plant development

Chapter 8

APPLYING TECHNOLOGY TO THE BOTTOM LINE IN A PRIVATISED ENVIRONMENT

A E White (Biwater Ltd, UK)

State-of-the-Art Engineering in the Public Sector has often become a necessity to preserve Pride of Place

A 25,000 m³/day treatment works, with similar influent and effluent, designed and constructed in the North of England can cost 50% of the cost of a plant of identical capacity in Southern England and vice versa.

The difference in cost in this hypothetical, but real, example is the approach by the Engineer/Architect. Leaving aside the cost of delay of a possible 12 months caused by a 'one off' design, it is still the case today that many projects are costly because they become monuments to the Engineer responsible. This approach is fostered and encouraged indirectly by our own Institution whose 'visit to a works' such as that proposed for this afternoon, can foster an excellence of construction and finish, coupled with the latest 'state of the art' technology, way out of commercial reason and justification.

The privatised, competitive environment will not permit Engineers to build bridges with supports at one metre centres in titanium

Consulting Engineers cannot escape similar criticism where over-engineered design means self protection beyond that required by technical engineering prudence.

Water and sewage treatment plant buildings, when designed by Architects, have, on occasions, emphasised standards of finish appropriate to new General Hospitals. This is not only unnecessary but unforgiveable especially when it is at the expense of mechanical equipment in the 'operating theatre'.

There will be few engineers here today who will not be able to point to examples in their own Authorities or Company where 'appearance and technical excellence' is out of step with commercial reality. Covering rapid gravity filters, tiling of pumping stations, and concrete in preference to steel must become cost effective as well as technical niceties.

Water Treatment – Proceedings of the 1st International Conference, pp. 69–74

New technology in a fiercely commercial environment will introduce value engineering on a new scale

We are now in a fiercely commercial environment not because we have a competitor, as the monopoly is still intact, but when a neighbouring Water Company Ltd is supplying its consumers with water at half the tariff of an Authority plc, questions will be asked not only by the consumers but by shareholders.

The introduction of a "fully automated works" or "total telemetry" will now only proceed if the 'rate of return' on the capital required can be justified over that for a manual or semi-manual operation. Can we, as Engineers, use our skills, not to provide that which is technically best but to provide a safe minimum of technical design that we can justify both in terms of safety and cost.

We have accepted plastic technology as a cheaper alternative to steel or iron but as engineers this still took some swallowing and has taken nearly 20 years to be fully accepted.

Ours is an old fashioned industry and change does not come easily especially where our consumers' health is at risk but today we need to lead with the implementation of new technology.

Capital needs to be conserved and its expenditure fully justified which should catapult our conservative industry into leading with new technology where capital or running costs can be safely cut.

Costs per capita will introduce financial engineering as the first criteria for designs which are required to meet only the minimum environmental standards of the day.

Exporting engineers and contractors have had to demonstrate talents for financial engineering to enable their clients to not only have the most attractive offer but often to enable their Clients to proceed at all.

Technology can enable financial engineering to give a third world client a more expensive plant at a cheaper price e.g.

The same Treatment Plant:

(a)	(b)
80% imported UK steel & machinery	20% imported UK steel & machinery
20% concrete	80% concrete
1 million cost	800,000 cost

Financed as follows:

	(a)	(b)
Grant Aid 35% of UK element	280,000	56,000
ECGD Guaranteed loan 65% of UK element	520,000	104,000
Local costs	200,000	640,000
	1,000,000	800,000
less Aid Grant	280,000	56,000
cost to client	720,000	744,000

The same balance of technology versus finance will be applied increasingly in the UK and eventually throughout Europe where a 'commercial return' versus the required environmental standard will become the order of the day. An example of this, but with non-concrete alternatives (on the grounds of cost and not finance) is as follows:

A one million gallon (4,500m^3) open tank (reservoir/settling tank/holding tank etc)

Concrete floor & glass-coated
        steel walls   £75,145
Concrete floor & concrete walls   £135,440

These examples are not high technology but meet their commercial and technical objectives at substantially lower cost.

Effluent quality has never before come under the microscope to the extent that is called for today and vast capital expenditure is required to bring EEC effluent standards up to those set by the EEC. It is therefore important not only from the capital cost criteria but for those of maintenance and operation that designs are standardised.

Basic technology can show a revival rather than a decline where the "man in a van" is cheaper for both plant operation and data recording than the cost of capital required to provide remote operation and data supply.

Conversely, despite the efficiency and economy of existing slow sand filters, for example, the value of the land used could be better applied by the use of higher rate filtration to show a nett capital gain to reduce future or existing borrowing.

New technology in low-head crossflow turbines have increased their use and viability on river weirs, night charging and outfalls:

Low Head Filtration. **The continual Automatic Backwashing Filter** represents new technology in the UK Water Industry as it is based on "Low Head" filtration. It combines a new approach to filtering, which dramatically reduces capital and running costs, with a modular dynamic backwashing system. The package is very efficient in regard to space utilisation; the structures are simple to construct and savings of up to 20% can be achieved when compared with conventional rapid gravity filters. The units are fully automatic and self-contained. Where manning levels are minimised, automation costs can escalate but not with this filter. Up to 50% savings can be realised by the use of a standardised Programmable Logic Controller that sequences the unit and is able to monitor performance. The low total power and low instantaneous backwash flow mean further savings in peripherals such as cabling, transformers and wash water handling. Similarly, because wash water is derived direct from the filtered water channel, there need be no washwater storage tank.

Pellet Reactors. **Softening** of bulk water supplies has always produced, as a by-product, large volumes of sludge that present significant problems of disposal. Traditional systems are also limited by low settling velocities or high replacement cost of resins. Both of these are overcome with pellet reactor softening - the immediate benefits being more efficient use of site area and an easily handled, self dewatering sludge (in the form of hard spherical pellets).

Because it is a high rate process (approximately 10 times faster than conventional lime systems) based on sand seeded fluidised bed principles, it is also easier to control and automate - this added robustness actually simplifies running and therefore allows economies in operating costs. Over the life of the plant, savings in the order of 10-15% can be expected - the larger the works the greater the proportional savings.

Hybrid Sewage Treatment. **Energy** efficient processes have been developed which reduce operating costs by up to 25% over conventional methods. Mechanical surface aeration is known to be more efficient over the early stages of the aeration process, whilst fine bubble diffused air systems have peak efficiency later in the aeration process when the basic polluting matter has begun to break down. By carefully combining the two processes into a hybrid system we can produce a plant in which the aeration process efficiency is continually maximised.

The Biwater Tower

Advances in water treatment and environmental management are, of course, utilised totally differently whether in Europe, or, say, the Third World.

The Biwater Tower is one example of an advance in water treatment which could have a substantial impact on Third World development. It streamlines a conventional water supply system by omitting double pumping and a clear water holding tank. The speed with which this advanced form of water treatment plant can be installed enables a complete State or Region to be completely serviced by piped, treated water within a period as short as 12 months. The "advance" enables fabrication both overseas or in the recipient market dependent on the availability of local steel plate or off-shore funding. To date nearly 200 of these Towers have been installed in West Africa and the Far East and are incorporated within projects totalling over $500 million. The smallest Tower, which can be erected in one day, can be supplied complete for approximately $25,000.

To over-simplify the benefits of the Tower, it can be said that it is a conventional reservoir with a water treatment plant supplied free of additional cost!

Where 2/3rds of the world are still without piped, treated water, this advance in water treatment will make a substantial impact because of its simplicity of operation as the whole unit's basic function is operated by one multi-port valve.

The Tower is not an advance in water technology as conventional sedimentation and filtration rates are used.

The environmental impact of the conventional reinforced concrete construction is such that it is difficult to remove or enlarge when obsolete, whereas, of course, steel can have an after-life and is easily removable.

The Biwater Tower is only one instance of dozens of new advances in both water and effluent treatment where engineering excellence has to be balanced with financial engineering to enable both the developed and the Third World to make piped, treated water affordable.

In conclusion

Profit now ranks in priority with water and effluent standards. The above examples of new technology are but a few of the hundreds of opportunities open to the design engineer to make every penny spent, count.

The true meaning of an 'engineer' is now back in fashion.

Chapter 9

THE DEVELOPMENT OF AN AERATED FILTER PACKAGE PLANT

A J Smith, J J Quinn and P J Hardy (Thames Water plc, Isleworth, UK)

ABSTRACT.

The Biological Aerated Filter is a comparatively recent development in sewage treatment. This high rate process comprises an aerated submerged packed bed of granular silicatic media with upward or downward wastewater flow.

The high specific surface area of the media allows the attachment of a high concentration of biomass, typically 4 times that found in suspended growth systems. As a result the process requires an aeration tank volume significantly smaller than conventional process options. In addition, the majority of suspended solids are removed in passing through the bed and final settlement is not required.

A pilot scale evaluation of an aerated treatment filter combined with an integral high rate polishing filter has been carried out for the complete treatment of settled sewage.

A good quality effluent with ammoniacal nitrogen concentrations less than 2mg/l was achieved at volumetric loading rates of 1.7kg BOD_5 /m^3.d and 0.6 kgTKN/m^3.d.

Using the results from the pilot plant a full scale prefabricated biological aerated filter package plant was designed and installed at Chesham sewage treatment works, containing aeration and filtration zones in a single module. Since commissioning the plant has reliably produced a good quality effluent.

1. INTRODUCTION

Increases in population and water use and improvements in effluent and river quality standards require the development of quick-build, cost effective solutions to uprate and upgrade existing sewage treatment plant.

Public awareness of both environmental and financial issues, and recent legislation has reinforced the need for high-rate, prefabricated, package treatment plant as alternatives to conventional process options. Most current sewage treatment options in the United Kingdom are based on aerobic processes in use since the early 1900's[1,2].

Development has led to increases in efficiency and quality, however large "aeration" tank volumes, final settlement and tertiary treatment are normally required to produce good quality fully nitrified effluents. As a result their use in prefabricated package plants is limited by size.

A comparatively recent development is the Biological Aerated Filter. First patented in North America in the early 1970's[3], a number of commercial variants are available (Biocarbone[4], Biofor[5]) and plants are operational in France, Japan and North America.

Work in the UK has been limited. A pilot plant evaluation for carbonaceous oxidation of settled sewage and tertiary nitrification of secondary effluent has been carried out by the Water Research Centre[6]. Further pilot scale work has been carried out by Severn Trent Water plc[7] into the production of high quality effluents and a large scale evaluation for complete treatment on a 1250 population equivalent plant has been carried out by Thames Water[8].

The process comprises a submerged packed bed of granular media. Air is introduced at or near the base with wastewater flowing upwards or downwards through the bed.

The large specific surface area of the media allows the growth of a high concentration of attached biomass which is responsible for treatment, typically 4 to 5 times that found in suspended growth systems. As a result the process requires aeration tank volumes significantly smaller than conventional process options.

Table 1 shows a comparison of 5 day Biochemical oxygen demand volumetric loading rates (VLR) for conventional treatment options and the biological aerated filter. For an equivalent effluent quality, the aerated filter requires an aeration tank volume approximately 3 times smaller than an activated sludge plant and 20 times smaller than a percolating filter.

In addition to the relatively small aeration tank size, suspended solids removal is accomplished throughout the bed and effluent quality is often good enough to discharge directly to the receiving watercourse without final settlement.

Where good quality effluents are required final effluent polishing can be achieved by a static portion of the bed below an intermediate aeration grid, as in the Biocarbone process, or by a separate high rate filter.

Initial work was carried out on a pilot plant at Thames Water's Ash Vale research facility. The objective of the study was to obtain

Table 1
Process Volumetric Loading Rates
for 90% Nitrification

	BOD_5
Percolating Filters	0.06
Plug Flow Activated Sludge[1]	0.42
High Rate Oxidation Ditch[2]	0.35
Low Rate Oxidation Ditch[3]	0.18
Biological Aerated Filter (BAF)	1.2 - 2.5[4]

Note: 1. Volumetric loading rate (VLR) $kgBOD_5/m^3media.d$
2. Food to microorganism ratio (F:M) 0.12
   Mixed liquor solids (MLSS) 3500 mg/l
3. F:M 0.1, MLSS 3500 mg/l
4. F:M 0.05, MLSS 3500 mg/l
5. Dependent on total nitrogen VLR

information on the performance of an aerated flooded filter with separate high rate solids removal stage. Using the process information gained from the pilot plant a full scale aerated filter package plant was designed and installed at Chesham Sewage Treatment works.

2. PLANT

A schematic of the aerated filter pilot plant is shown in Fig.1. The plant is based on a steel container housing 4 1m x 1m x 4m high filter cells with an integral 4m x 1m x 3m high washwater reservoir. Inlet and outlet connections are provided to allow each filter to be operated independently, in parallel or series, as either aerated treatment units or as an upflow or downflow polishing filter.
Process air was supplied from a coarse bubble aeration grid positioned on the false floor above the plenum chamber of each treatment filter.
Washwater was supplied through 4 separate submersible pumps mounted in the washwater reservoir and connected directly to the washwater inlet of each plenum chamber. The backwash air scour was supplied by a single blower mounted within the control section of the container. Backwash liquors were removed from the top of each filter by submersible pumps and returned to the main works inlet. Valves were

electrically actuated by a computerised control system. The backwash cycle could be initiated manually or automatically on time or headloss with sequence control by contact electrodes.

A number of different support media types were evaluated for hardness, biomass attachement and headloss development. The media chosen comprised a silicate material with a rough porous structure and a 3mm effective grain size. Both treatment and polishing filters

FIGURE 1.

SCHEMATIC OF AERATED FILTER PILOT PLANT — ASH VALE

1. PLANT DISCHARGE, AIR SCOUR & LEVEL CONTROL TANK SYSTEMS

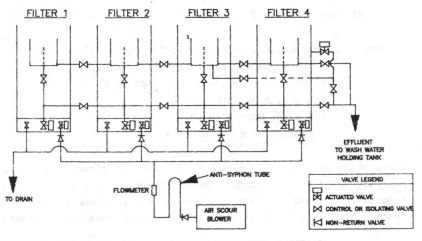

2. PLANT FEED, PROCESS AIR & BACKWASH WATER SYSTEMS

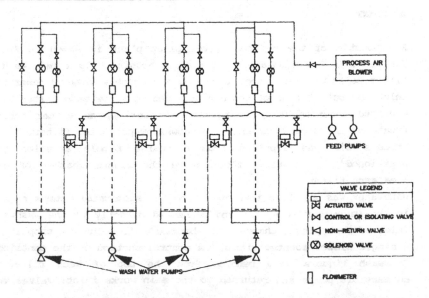

were filled with the same media to depths of 2m and 1.8m respectively.

Screened, degritted raw sewage from the main works flow was fed to a primary sedimentation tank. Settled sewage flowed into a holding tank and was pumped from there to the inlet of each treatment filter.

3.METHOD

The programme of work for the pilot plant was divided into 2 operational phases. During the first period the plant was operated with 3 downflow aerated filters treating the same load with equal air supply, followed by a downflow polishing filter. In the second period the load to 2 of the treatment filters was increased, with the third acting as a control on unchanged load.

Early results from the pilot plant evaluation showed the feasibility of larger sized plant based on an aerated treatment zone followed by an integral high rate polishing filter.

As a result a full scale treatment module was designed and installed at Chesham sewage treatment works. The performance of the new plant was monitored over a period of months.

Samples of influent, treatment filter effluent and polishing filter effluent were taken from the pilot plant, with influent and effluent samples from the full-scale plant. The samples were analysed for suspended solids (SS), 5 day Allyl Thiourea inhibited biochemical oxygen demand (BOD_5), total kjeldahl nitrogen (TKN), ammoniacal nitrogen (NH_4-N), nitrite nitrogen (NO_2-N) and nitrate-nitrogen (NO_3-N).

4.RESULTS

Start up of the 3 treatment filters was achieved by initial batch aeration of settled sewage followed by a period of continuous low flow. The flow was increased from 5.5 l/m to 16.7 l/m over a 1 month period.

During the steady state phase the flow to the treatment filters was kept constant at 16.7 l/m. All of the flow from the 3 treatment filters was passed to the downflow polishing filter giving a total flow of 50.1 l/m. These flows relate to hydraulic loading rates (HLRs) of 1.0 m/h and 3.0 m/h for treatment and filtration respectively.

The process air flow rate was kept constant at a rate calculated from the stoichiometric oxygen demand for carbonaceous oxidation and nitrification and an assumed oxygen transfer efficiency.

The average volumetric loading rates (VLR) for this period were 0.86 kg BOD_5/m^3.d, 0.18 kg NH_4-N/m^3.d and 0.36 kg TKN/m^3.d Effluent

FIGURE 2

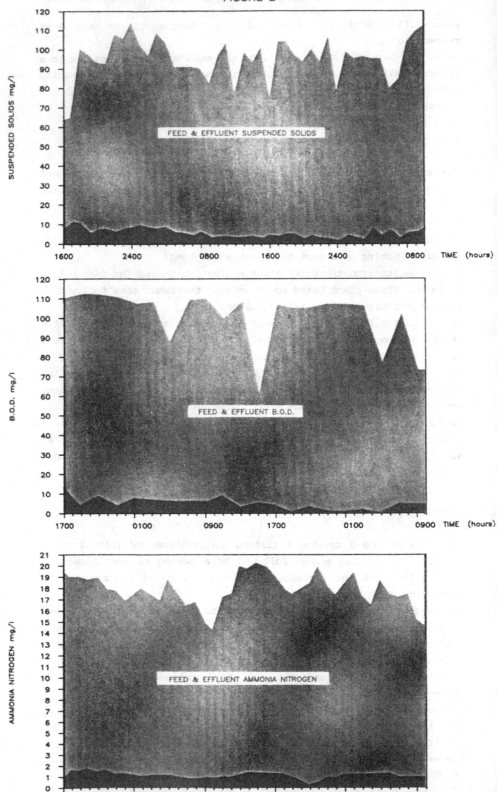

quality averaged 3.9 mg/l SS, 3.6 mg/l BOD_5 and 1.6 mg/l NH_4-N from the base of the treatment filters.

The final effluent from the polishing filter averaged 2.2 mg/l SS, 2.1 mg/l BOD_5 and 0.7mg/l NH_4-N, equivalent to removal rates of 98% SS, 97% BOD_5 and 95% NH_4-N.

Although on constant flow the the settled sewage strength varied throughout the day subjecting the plant to partial diurnal load variation. A number of intensive surveys were carried out by analysing individual 24 hourly samples of influent and effluent. Typical results for Filter 1 are shown in Fig.2.

Backwashing of the filters was carried out every 24 hours using a cycled wash sequence involving periods of air scour, combined air and water, high rate water and a pumped drain down. The volume of effluent used daily for backwashing was equivalent to 8% of the total flow.

Effluent quality from the treatment filters deteriorated immediately after backwashing. Typically, suspended solids rose to a maximum of 60mg/l and returned to less than 20mg/l 45 minutes after backwashing, as illustrated in Fig.3.

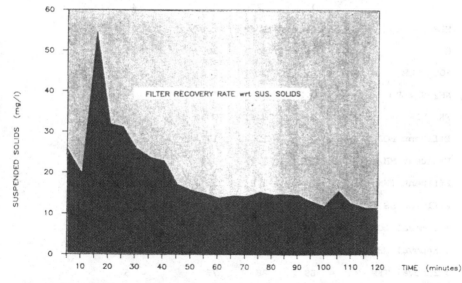

Final effluent quality was not affected during this period because of the action of the polishing filter in removing the extra solids and the dilution effect of the other filters still in operation. During the load variation phase the sewage flow to filters 2 and 3 was increased in 2 stages. The flow to filter 1 was kept constant as a control. Each filter was allowed to reach equilibrium over a 7 day period before sampling. Each operating period lasted approximately 21 days. Process air flow to all filters was kept constant.

Filter 1 continued to produce a good quality effluent with over 90% BOD_5, SS and NH_4-N removal at VLR's of 1.2-1.3 kg $BOD_5/m^3/d$ and 0.4-0.45 kg $TKN/m^3/d$. Effluent quality averaged 7 mg/l BOD_5, 8 mg/l SS and 1.3 mg/l NH_4-N. Nitrification remained close to 90% up to the limiting TKN VLR of 0.6 $kg/m^3.d$.

Above this figure a reduction in NH_4-N removal efficiency was seen. The maximum VLR's of 3.85 $kgBOD_5/m^3.d$ and 1.35 $kgTKN/m^3.d$ gave 75% BOD removal and 69% solids removal. Nitrification, as indicated by both NH_4-N removal and NO_3-N formation, was negligible however a 22% reduction in TKN was observed.

Operating data and plant performance for the 2 periods are given in Tables II and III.

Table II

Load Variation Period 1

Filter Number

	Aeration			Filtration
	1	2	3	4
HLR	1.0	1.5	2.0	4.5
HRT	60	40	30	13
BOD_5 VLR	1.2	1.7	2.3	-
NH_4-N VLR	0.2	0.3	0.4	-
TKN VLR	0.4	0.6	0.8	-
Effluent BOD_5	7	12	17	8
Effluent NH_4-N	1.1	1.7	8.4	2.6
Effluent TKN	3.7	3.8	8.3	6.2
Effluent SS	5	7	13	4
% removal BOD_5	93	88	82	38
% removal NH_4-N	93	89	47	42
% removal TKN	89	88	75	6
% removal SS	95	93	87	58

Note: HLR : Hydraulic loading rate (m/h)
         HRT : Hydraulic retention time (based on 50% voidage) (min)
         VLR : Volumetric loading rate $kg/m^3media.d$

Table III

Load Variation Period 2

Filter Number

	Aeration			Filtration
	1	2	3	4
HLR	1	2	3	6
HRT	60	30	20	10
BOD$_5$-VLR	1.3	2.6	3.9	–
NH$_4$N-VLR	0.2	0.4	0.6	–
TKN-VLR	0.45	0.9	1.35	–
Effluent BOD$_5$	7	13	17	11
Effluent NH$_4$-N	1.5	7.9	13.1	7.7
Effluent TKN	7	18	25	18
Effluent SS	10	17	32	12
% removal BOD$_5$	93	91	75	42
% removal NH$_4$-N	91	52	2	18
% removal TKN	78	45	22	8
% removal SS	90	83	69	48

Note: HLR : Hydraulic loading rate (m/h)
       HRT : Hydraulic retention time (based on 50% voidage) (min)
       VLR : Volumetric loading rate kg/m^{3}media.d

5. PACKAGE PLANT DESIGN

Using the data from the pilot plant evaluation a large scale package
aerated filter plant was designed and installed at Chesham sewage
treatment works. The plant was based on prefabricated relocatable
steel containers for treatment, washwater storage and pumping, plant
control, and process and backwash air supply.

The overall dimensions of each module were governed by the
constraints imposed by transport to site however, an increase in
treatment capacity is possible simply by increasing the number of
treatment units as required.

The treatment module comprises an aerated treatment section, high
rate downflow polishing filter and integral pipework and instrument
housing. A schematic is shown in Fig.4.

The tank was constructed from 6mm thick mild steel plate with
appropriate stiffening to allow it to be moved full of media. Based

on maximum downflow velocities a surface area ratio of 6:1
treatment:filtration was used in design. The aerated section has
36m^3 of expanded shale media, with 4.5m^3 of the same media in the
polishing filter.

The floor of both filters is made up of 1m^2 galvanised steel plates
holding the filter nozzles. Process air is supplied by an aeration
system mounted on the floor of the treatment zone.

Pipework, actuated valves and flow measurement of settled sewage,
process air, backwash air and wash water are mounted in a housing
attached to one end of the module.

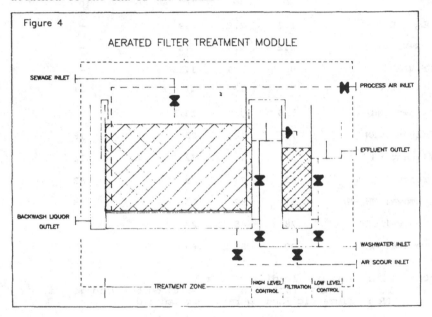

Figure 4

AERATED FILTER TREATMENT MODULE

6.OPERATION

Settled sewage is pumped from an existing sump to the inlet in the
top of the treatment section and flows down through the aerated
media into the plenum chamber. Effluent from the plenum chamber
flows up over a high level weir, which maintains the liquid level
in the treatment filter and into the top of the polishing filter.
The flow passes down through the polishing filter where excess
solids are removed. Final effluent flows into the washwater storage
reservoir and out into the receiving watercourse. Backwashing of
both treatment and polishing filters is carried out simultaneously
under automatic control. The backwash sequence involves periods of
air scour only, air scour plus low rate water wash and high rate
wash, as in the pilot plant.

Although designed as a research and development installation the
plant was required to treat a constant 1000 m^3/d offload from the
main works to a 7mg/l 95 percentile standard while parts of the

existing works are demolished to make way for a new oxidation ditch. Since commissioning the plant has consistantly produced a good quality effluent averaging 13 mg/lSS, 14 mg/lBOD$_5$ and 3.8 mg/lNH$_4$-N. Average VLR's for the same period were 2.2 kgBOD$_5$/m^3.d, 0.37 kgNH$_4$-N/m^3.d and 0.68 kgTKN/m^3.d. The calculation of VLR was based on the volume of media under aeration.

Operating data for November 1989 is given in Table IV. Examination of the media has shown an attached growth of small, thin rod bacteria. Examination of the backwash liquors has shown a predominance of Sphaerotilus spp., Thiothrix spp. and Beggiotoa sp. with colonies of Opercularia spp., Carchesium spp. and Vorticella spp. The predominant species are often associated with low dissolved oxygen levels or nutrient defficiency[9,10].

Table IV
Chesham Aerated Filter Package Plant

Operating Data November 1989

	Av	Max	Min
HLR	1.8	2.2	1.6
HRT	31	38	26
BOD$_5$VLR	2.2	3.1	1.5
NH$_4$-NVLR	0.37	0.5	0.27
TKN-VLR	0.68	0.81	0.48
Effluent BOD$_5$	14	19	11
Effluent NH$_4$-N	3.8	6.9	0.8
Effluent TKN	14.3	16.1	12.0
Effluent SS	13.0	13.5	12.0
% removal BOD$_5$	88	90	86
% removal NH$_4$-N	78	94	57
% removal TKN	58	66	48
% removal SS	82	87	76

Note: HLR : Hydraulic loading rate (m/h)
       HRT : Hydraulic retention time (based on 50% voidage) (min)
       VLR : Volumetric loading rate kg/m^{3}media.d

The characteristic grey filamentous growth of Beggiotoa sp. and Sphaerotilus spp. were also found on the outlet weir of the treatment unit although dissolved oxygen levels in the effluent were normaly greater than 2mg/l.

The supernatant liquid above the bed was similar to the backwash liquor although lower in suspended solids. At certain times, however it contained a suspended growth of large light coloured flocs. These flocs comprised stellate clumps of filaments intermeshed with a mass of rod shaped bacteria.

The appearance of these flocs corresponded with periods of exceptional effluent clarity although no link with any process or climatic variable was found.

Backwashing of the filter was carried out every 24 hours or on reaching a predetermined loss of head. The volume of washwater used was between 5% and 8% of the average daily flow.

7. CONCLUSIONS

The results from the pilot evaluation and full scale plant show that biological aerated filters can reliably produce good quality nitrified effluents, providing a high rate alternative to conventional treatment options.

Effluent quality was governed by organic, nitrogen and hydraulic loading rates. Pilot plant effluent NH_4-N concentrations remained below 2mg/l up to VLR's of 0.6 Kg TKN/m^3.d. and 1.7 Kg BOD_5/m^3.d., with a residence time of only 40 minutes. Effluent from the full scale package plant averaged 13 mg/l SS, 14 mg/l BOD_5, and 3.8 mg/l NH_4-N at VLR's of 2.2 $kgBOD_5/m^3$.d and 0.68 $kgTKN/m^3$d.

High concentrations of attached biomass resulted in comparatively small aeration tank volumes and no final settlement was required, making the process ideal for prefabricated package installations.

Although filamentous organisms were present in and above the bed no adverse effect on effluent quality was observed.

Treatment filter effluent suspended solids increased with increasing hydraulic and organic loading rates. The integral high-rate tertiary filtration stage and dilution effect from other filters ensured a good quality final effluent at all times. With less stringent SS and BOD_5 standards the dilution effect alone could ensure the required effluent quality.

The separation of treatment and filtration allowed a greater aerated bed depth to maximise nitrification and optimise final solids removal. The use of prefabricated steel plant minimises site work and reduces the overall construction period. Initial estimates have shown a saving in capital costs over conventional process and construction options. In addition, changes in load or effluent quality requirements can be catered for by increasing or decreasing the number of treatment modules.

ACKNOWLEDGEMENTS.

The work was carried out as part of the Thames Water plc Research and Development Programme. Thanks are due to Wayne Edwards, Chris

Woods and Karen Aston for plant operation and microbiological analysis. Any views expressed in this paper are those of the authors and not necessarily those of Thames Water plc.

REFERENCES.

1. Ardern, E. and Lockett, W.T., 1914, " Experiments of the Oxidation of Sewage without the Aid of Filters, Parts I, II and III" J.Soc.Chem.Ind., 33, 523.

2. Royal Commission on Sewage Disposal, 1908, Fifth Report, H.M. Stationary Office.

3. Biological Aerated Filters. Canadian Patent No. 953039, 1974.

4. Leglise, J.P., Gilles, P., and Moreaud, H., 1980," A new Development in Biological Aerated Filters." Presented at the 53rd Annual WPCF Conference, Las Vegas.

5. Amar, D., Faup, G.M., Richard, Y., and Partos, J., 1984, "Epuration aerobie par cultures fixees: Procede Biofor." Proceedings of the 7th Symposium on Wastewater Treatment, Montreal, Canada.

6. Dillon, G.R., and Thomas, V.K., 1989, " A Pilot Scale Evaluation of the Biocarbone Process for the Treatment of Settled Sewage and for Tertiary Nitrification of Secondary Effluent." Proceedings of the IWAPRC/EWPCA Conference on Technological Advances in Biofilm Reactors, Nice, France.

7. Lilly, W., Bourn, G., Crabtree, H., Upton, J., and Thomas, V., " The production of high quality effluents in sewage treatment using the Biocarbone Process." Paper presented Inst.Wat.Env.Man., West Midlands Branch, October 1989.

8. Thames Water, 1989, " Biocarbone Process Evaluation", Research and Development Internal Report.

9. Richard, M., Hao, O., and Jenkins, D.,1985," Growth Kinetics of Sphaerotilus Species and their significance in Activated Sludge." J.Water Pollut. Control Fed., 57, 68.

10. Hattingh, W.H.J.,1963," The Nitrogen and Phosphorous requirements of Micro-organisms." Water Waste Treat., 10, 380.

Chapter 10

PLANT MONITORING AND CONTROL SYSTEMS: MAKING THE CASE AND MANAGING THE IMPLEMENTATION

M F Williams (Severn Trent Water, UK), D L Gartside (Kennedy & Donkin Systems Control, UK)

ABSTRACT: Severn Trent Water has in excess of 5000 operational sites which require the monitoring and control of mechanical and electrical plant.

In late 1988 a study was made of the plant monitoring and control issues which resulted in a business case for a telemetry system to provide remote monitoring and control of 3500 sites. Design and implementation of the system commenced in early 1989 with a target for completion by the end of 1992.

The Paper describes the business case methodology and the project management techniques used to define, design and implement the system.

The Business Case preparation included a review of appropriate design solutions, cost estimates, interfacing of existing telemetry and ICA systems, implementation and programming issues.

This involved use of sample engineering audits of operational sites to determine site functionality, existing investment in ICA and Communications. Overall scheme requirements being extrapolated using databases and financial modelling techniques.

In the design of the scheme, users requirements were satisfied by producing design intent documents for Telemetry, Instrumentation, Communication and interfaces to ICA Schemes. Presentations and demonstrations to users were a key feature to obtaining user acceptance.

Water Treatment – Proceedings of the 1st International Conference, pp. 89–104

Model specifications and Standards applicable to the scheme and other related projects were produced to fast track the implementation and to maximise existing capital investment on scheme sites.

The siteworks involved close coordination between the systems engineers and operations personnel to enable speedy, efficient and progressive completion of activities.

INTRODUCTION

Severn Trent Water provide vital water services to a population of over 8 million people in over 20 thousand square kilometers in the centre of England. Severn Trent Water own of the order of 800 reservoirs, 1000 sewage works, 40 major water treatment works and over 70 thousand kilometers of underground mains and sewers. The business has an annual turnover of approximately 500 million pounds, an annual capital investment of over 160 million pounds and employs approximately 8 thousand people.

Severn Trent Water is responsible for managing an inheritance of a fragmented, complex network of water supply systems, sewage and effluent treatment plants and other associated resources, some of which are over one hundred years old. The organisational structure of Severn Trent Water has evolved to support the changing needs of the business. The business is operated on firm commercial principles with all new investment justified against clearly defined objectives.

Experience of the application of technology to the business has, in the past, not been entirely successful. Therefore, to meet the challenge of a major new initiative for the improved operation of Severn Trent Water's resources, a detailed business case was prepared by a project team comprising Severn Trent Water staff, Kennedy and Donkin Systems Control engineering consultants and a management consultant. The successful business case is the basis of an ambitious but realistic programme for the implementation of a Plant Monitoring and Control System within Severn Trent Water.

The paper describes the background to the initiative, the methodology employed for the business case and the design and implementation philosophy for the Plant Monitoring and Control System.

BACKGROUND

The organisational structure and management of Severn Trent Water has changed considerably in the transformation from local authority control to that of a private water company. Such changes have included staff reductions, a variety of organisational structures and increasing concern about customer expectations and environmental legislation.

Against this background of change, the 1988 Severn Trent Water business plan states 'The main objectives of the organisation are to improve our services where necessary and maintain them over most of the region where they are satisfactory'.

One of the many ways Severn Trent Water expects to maintain the trend of lower operating costs is to 'Ensure that organisational structure is appropriate for the services to be provided in the light of improving technology and techniques'.

Technology in the form of telemetry has contributed to the downward trend of operating costs whilst upgrading the level of service through an increased operational monitoring and control capability.

The present organisational structure provides a clear definition between operations and support services. Fifteen operational districts have been formed each managed by an accountable District Manager. District Managers inherited a variety of instrumentation, control and telemetry systems with a wide variance in age, capability and usefulness. Seven of the 15 operating districts relied on systems which were perceived to be giving reduced reliability and increased cost of maintenance.

Equally Severn Trent Water were aware, and had experienced the pitfalls and problems where technology rather than the operational need, determined the levels of investment. Also, within the wider ICA industry, many cases are reported of project failures due to operator dissatisfaction, increased cost of ownership, inappropriate and often over-engineered design solutions, extended development programmes and protracted commissioning.

Severn Trent Water adopted a pragmatic and business approach to resolving the issue of what, if anything, was required to maintain and improve the supervision and management of district sites and assets. The principal questions asked were:

(i)   What are the available solution options?

(ii)   How do they compare?

(iii)   What is the optimum level of investment commensurate with the benefits to be obtained?

In September 1988, Severn Trent Water formed a small project team to address the above issues. This team comprised Severn Trent Water engineering services, Kennedy and Donkin Systems Control Ltd (KDSC) as engineering consultants and a management consultant. The project (if indeed there was a need for one), was to be titled 'The Plant Monitoring and Control System (PMCS)'.

The project team's first task was to present the first stage business case report to the Severn Trent Board of Directors.

THE BUSINESS CASE

The core of the business case approach is its emphasis on identifying and achieving business objectives. A significant element in the approach is the concentration on operational management and recognition of the key role of District Managers as owners and users.

Each District Manager has a responsibility to:

(i)   utilise their staff (and others as necessary) to define the need,

(ii)   identify the achievable benefits,

(iii)   directly support the provision of the eventual solution.

Considerable effort was expended to change existing cultural and organisational attitudes to this more rigorous and 'fast-track' approach.

Lead by the Severn Trent Water PMCS Project Manager the team addressed the following areas.

The definition of need and opportunity, the project description, the position of the project in relation to the water business and activities, the relationship with and interface to other capital projects or initiatives and the relevance to the business plan. This report was completed within 2 weeks and submitted to the sanctioning board on the 21st September 1988.

Running in parallel with the above activity assessments of potential benefits, design solution options were studied and an assessment of risks leading to a recommended solution was made. This report was complete within 5 weeks and submitted to the sanctioning board on the 13th October 1988.

Upon the sanctioning board's agreement to proceed, and working within the budget determined and allocated at this stage, the next task was to assess in detail the first stage of the implementation.

This assessment addressed the detailed benefits associated with and accounted to each District Manager, detail costs and risk assessments, project timescale and cash flow projections.

The recommendation in stage 1 was for a phased implementation across the three areas based on need. Southern Area had the greatest need and was to be implemented first, the North and East need was almost as great and was to follow immediately after the Southern approval. The Western area need was not great and was not scheduled to be started until the other two area systems were operational. See Figure 1 and 2.

The report for the Southern Area of Severn Trent Water was submitted to the sanctioning board on the 11th November 1988 and approval to proceed was issued on the same day.

FIGURE 1 — STW — SOUTHERN PMCS

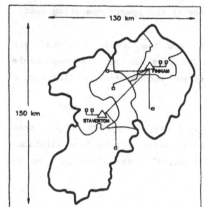

KEY STATISTICS

5	OPERATIONAL DISTRICTS
1100	SITES
£8.4M	CAPITAL VALUE
4.0M	POPULATION
11	LARGE WTW >20 Ml/d
12	LARGE STW >20 Ml/d
167	W.P.S.
645	S.P.S.
157	RESERVOIRS

DESIGN FEATURES

2	TELEMETRY CONTROL CENTRES
14	WORKSTATIONS AT DOC'S
15	WORKSTATIONS AT AGENCIES
23	MAJOR WORKS CONTROLLERS— DECISION SUPPORT/MIS
4–5	RTU DESIGNS

3 YEARS STRATEGY TO OPERATIONAL SYSTEM

FIGURE 2 — STW — NORTH EASTERN PMCS

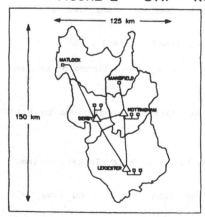

KEY STATISTICS

5	OPERATIONAL DISTRICTS
1250	SITES
£10M	CAPITAL VALUE
2.5M	POPULATION
8	LARGE WTW >20 Ml/d
10	LARGE STW >20 Ml/d
170	W.P.S.
660	S.P.S.
215	RESERVOIRS
36	BOREHOLE SITES

DESIGN FEATURES

3	TELEMETRY CONTROL CENTRES
16	WORKSTATIONS AT DOC'S
15	WORKSTATIONS AT AGENCIES
18	MAJOR WORKS CONTROLLERS—
5	RTU DESIGNS

SUPPORT/MIS

ANALYSIS OF NEEDS

In the business case approach the needs analysis focused on the operational need rather than the desirability of providing a facility or feature.

The Severn Trent Water business plan clearly identified the framework within which operational needs were to be determined by defining two specific objectives viz:

'Meet quality and service level specifications at minimum cost'.

'Remove need for further cost increases and, where possible, reduce costs further'.

Expenditure was to be applied where the greatest step in identified benefit was possible. The preferred solution was to be appraised by its ability to achieve a given level of service at the lowest operating cost.

This implied careful selection of sites and the level of signal and control provision for each site. Each District Manager was responsible for providing such information from often incomplete and inaccurate records.

To assist in the selection and definition of need, the PMCS project team had considerable data available to them covering similar installations throughout the UK and some very accurate detail information on some 80 Severn Trent Water sites.

This data, held on computer databases, comprised the following typical fields: site types, levels of signal provision, operational criticality, level of existing investments and benefit potential.

For PMCS, at the site level, the operational need was classified as:

(a)   Alarm only - the occurrence of a defined abnormal event.

(b)   Monitoring - the near constant (hourly) surveillance of process and plant conditions.

(c)   Remote control - the ability to start/stop operations on site from a remote location.

To assist in the evaluation of solutions, two service levels were defined:

(i)   Service levels as provided by existing telemetry on some 1900 sites.

(ii)   Service levels achieved on (i) provided for 3450 key sites.

In addition to the site related need the analysis determined the operational functionality required from any technology based solution. For example the type of displays and reports, the archiving of data and the need for operator assist facilities. Other important issues determined at this stage were the use of shared facilities, ownership and maintenance, reliability, resilience and performance requirements.

In summary, the needs assessment considered some 5600 in number sites in steps of 20, 40 and 70% covering 70 installation types. The level of provision ranged from simple alarm monitoring to remote monitoring and control across a mix of installation types. Against operational need the following opportunities were assessed; reduced manpower and transport costs, reduced electricity costs, improved asset management information, improved customer service and improved business management information.

DESIGN SOLUTION OPTIONS

Following on from the needs analysis, three solution approaches were pursued:

    (a)   do nothing
    (b)   a manual solution
    (c)   a technical solution.

The 'do nothing' solution was rejected early in the project as the solution did not allow existing service levels to be maintained.

The 'manual' solution assumed that all expenditure on existing telemetry would cease and manpower would be employed to take over the functions of the existing telemetry systems as and when these systems fail.

The 'technical' solution involved the provision of telemetry, instrumentation and site wiring. To address the issue of scale and functionality the following technical solutions were considered:

(i)   Like for like replacement.

(ii)   Simple alarm monitoring of some 20% of sites in each district.

(iii)   District based systems capable of a mixed sole functionality over a range 20, 40 and 70% of sites.

(iv)   Area based systems similar to (iii) but with the additional capability of re-routing data between operational centres.

(v)   A regional system using shared equipment, organisational
         structure independent and providing a central service to
         district managers.

Within the business case methodology the emphasis on cost versus benefits
required that solutions were rigorously appraised with cost accuracies
within 10-15%.

From experience of similar schemes, the PMCS project team realised that
some 60-70% of the scheme cost would be committed to site related
activities, viz. signals, communications, type of outstation. This being
the very area where existing data was at its most incomplete.

Various telemetry solutions were drafted using models derived from within
the UK water industry and elsewhere. The specification of outstation
functionality and input/output size was determined using reference
database data of site types and likely signal provision and performing
statistical analysis to extract PMCS specific information. This along
with other statistically derived data was then subjected to pilot site
audits to test the analysis.

The computers and workstations components of the various designs were
again drafted from computer stored records of manufacturers' quoted and
recently tendered prices. An important element in all designs was the
identification of the need for software development. Clearly some
software modification and development may have been required, but the
impact of such development was to be clearly defined and the impact on
implementation determined.

Communications are a key component of all the technical solutions
considered. Based on the needs analysis a range of potential solutions
was considered.

Consistent with most UK schemes, the use of the British Telecom Private
Switched Telephone Network, already establshed in 90% of outstation to
telemetry computer applications was to be the preferred choice. For the
remaining communication lines to sites private wire and radio
communication were to be specified. The high level communications were
to be based on the manufacturer independent protocols supported
throughout Europe which offered high resilience and 'open' interconnection.

The solution options available for site investments required
consideration of existing asset life and reliability, the adequacy and
maintenance of existing assets and the suitability of measurement
equipment to deliver the operational need.

Generally the PMCS project did not attempt to replace existing compatible instruments, provide local automation on works or provide quality monitoring other than by interfacing to instruments which were current and necessary.

A major factor associated with site related solutions was the recognition that sites would need to continue to remain operational during the installation period and that existing instrumentation and telemetry would remain in service until the new facility had been proven operationally.

Considering some 3450 sites with a possible input/output count of between 10 to 50 signals on average, (developed from existing and new instruments) each with an instrument loop possibly requiring between 2 and 8 elements of work to be costed, the range of likely solutions available was considerable. Such a range of solutions can only be adequately handled by comprehensive databases and computational methods supported by an up to date cost database.

The utilisation of such a database is at this stage of the project a vital factor; providing the ability to manipulate various solution options within a short timescale. Equally, by adopting the correct database structure, the later stages of design and implementation are considerably enhanced by the provision of site related data and costs to the project team and senior management. See Figure 3.

FIGURE 3 – KDSC PROJECT DATA–BASE STRUCTURE

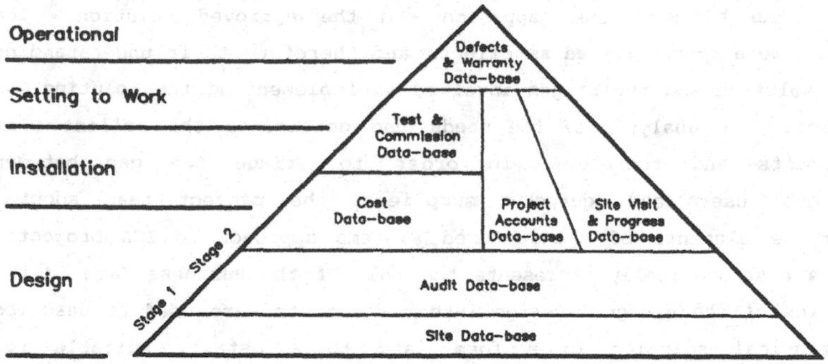

The assessment of technical solutions centred on the three most viable options viz. the district systems, the area systems and a regional system. Each option met the specification for the two service levels .

Each solution was evaluated against the following criteria:

(i)   what did it provide?

(ii)   the capital and revenue costs,

(iii)   risks to implementation including owner acceptance,

(iv)   potential benefits in cash terms,

(v)   major disadvantages.

Based on the assessment against the criteria a recommendation was made and accepted to adopt the solution based on area base telemetry systems.

The accepted solution was formed around three telemetry areas viz: Southern, North Eastern and Western which approximately coincided with the regional water resource system boundaries. Across the region some 3450 sites would be provided with alarm monitoring and control functionality and reporting to 16 operational centres.

DESIGN METHODOLOGY

The design phase had a number of key objectives. First the District owners had to perceive that they were the client and that the design was consistent with their expectations. Secondly the design activities had to define the details of the solution to enable costs and risk to be identified and defined. Thirdly the design activities had to enable suppliers to offer a compliant and competitively priced solution to the specifications. Design commenced in January 1989 and started with a series of presentations to the staff of the five Districts within the Southern Area. The emphasis of these early presentations was to report back on the business case approach and the approved solution. The districts were to be treated as clients and therefore their understanding of the solution and the stages involved in implementing the solution was paramount. The analysis of the needs had determined the salient user requirements and therefore, in order to bridge the gap between operational users and equipment suppliers, the project team adopted appropriate elements of a structured systems approach to ICA projects. Figure 4 diagramatically represents the role of the Business Case in the evaluation of the design. Design intent documents were used to describe the technical solution in a form, and to a detail, suitable for operational staff understanding and sanction.

In this way an adequate understanding of what was to be delivered was generated early in the project. Great stress was placed on a progressive appreciation and user acceptance of the solution.

Presentations were made to the district owners in June of 1989 and comprised formal presentations and demonstrations of typical man-machine interfaces. They also highlighted the issues of resilience and failures using computer models and simulations.

FIGURE 4 — DESIGN EVOLUTION AND ROLE OF THE BUSINESS CASE

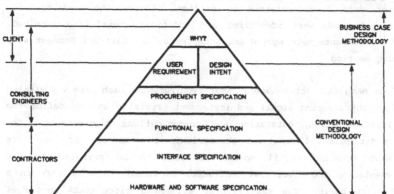

The design intent documents represent the highest level of system specification. The detailed specifications were generated using story board techniques and work shop sessions concentrating on specific areas of concern to the user. Issues such as flexibility, expandability, capacity and performance were addressed and specified as well as the need for operational data to be passed on to management information systems and the use of operational support computing within a telemetry network.

It was recognised that site work would consume a considerable amount of the scheme budget. During the preparation of the business case the level of signal provision for each potential benefit had been produced. This level of provision and the related scopes of work for each type of role were held in a database which associated a budgetary cost to each scope of work. The detailed signal provision formed the basis for determining the site by site, district by district user requirement specification.

Additional design intent documents were produced for instrumentation and site facilities such as communications and power. Instrument specifications and standards were prepared to define the details of work to be undertaken under site contracts. These standards are based on project experience and technical expertise which give the essential operational context to both the conditions found on site and the limitations of the possible instrument technologies and practices offered by contractors.

Within each district, the total number of sites were divided into planning groups, typically 4-5 in number representing 30-40 sites per group.

The design intent documents, standards and specifications influenced and guided the planning of site audits and the detailed definition of user requirements for each site.

User requirements were produced automatically for each site using the computer data bases referred to earlier. Changes to meet specific operational needs were identified and additional benefits determined. These requirements were agreed and accepted by the District Manager at a planning meeting.

Based on each District Managers' user requirements, each site was visited to audit the existing signal and instrument provision and to define the new instrumentation, outstation and communications requirements. The site audit provided data and photographic records of all the site equipment enabling detail scopes of work to be prepared for an instrumentation and electrical contractor to tender for the work on a fixed price basis. The same process allowed a close check on budget costings to reflect backwards to the business case estimate and forward to predict the likely outturn costs.

The mass of data generated from auditing the sites was assembled into the site database. A series of programs were run on the database computer to generate detailed site by site, loop by loop scopes of work, along with compilations of input/output lists necessary to size telemetry outstations, identify operational access constraints and define work items to be done by others eg. applications for the right of access.

The site database generated structured project reports presenting information on a site-by-site basis, and a summary of estimated costs at project, site and loop level so that Severn Trent Water could review and approve the design. Following client approval the site work tender specification and schedules were automatically generated from the databases.

PROCUREMENT AND IMPLEMENTATION

In parallel with the detailed definition of the requirements and the specification of the solution, the appropriate means of procurement needed to be addressed to ensure that the solution was delivered to the specified quality, cost and timescales.

Inherent in this stage was a need to minimise implementation risk and thereby secure a system capable of achieving the benefits. The procurement and implementation strategy recognised the technology split between telemetry and site work, that site work was to be carried out while sites remained operational, the large number of sites spread geographically and the similar and repetitive nature of many of the tasks and activities.

Based on these considerations, the telemetry systems (including outstations) were to be procured and managed as a single contract per area (viz. Southern, North Eastern and Western) and the site works were to be divided into some 14 contracts, each District based.

For the telemetry system the suppliers invited to bid were prequalified from the response to a public invitation, both in the UK and mainland Europe, to express interest. The pre-qualifications included a first stage questionnaire identifying broad technical and commercial suitability. This was followed by a second more rigorous review of technical capability and quality systems which involved visits to the candidate's premises and inspection of their practices and procedures. Also client and credit references were taken up from the potential suppliers.

Pre-qualification resulted in a tender list of five and a tender specification for the Southern Telemetry Area (STA) PMCS Telemetry System was released in August 1989 with a period of 2 months for response.

Evaluation of the five received tenders involved assessment of the technical compliance together with the analysis of the financial implications of various levels of definite and optional work. These levels were defined by the site audits completed to date, and schedules of anticipated procurement to be clarified during the remainder of the audit programme. To compare these aspects of the bids a number of models of the final priced configuration were established ranging from the minimum expected to the maximum anticipated expansion. These models took into account the cost of ownership which included maintenance software licences and expansion costs. Within this range the mix of outstation types was varied to reflect the uncertainty in the outcome of the incomplete audit activities. This process allowed the bids to be compared, costs assessed and the implications of the costs on the procurement strategy to be evaluated.

Arising from the telemetry specification and the tender returns for the Southern Area, the technical solution offered is of high quality providing a system able to meet the needs of the District Managers. Potential exists with the technical solution for expansion and enhancements without major implications to the PMCS investment. The system provides a variety of functionality in outstation types including the use of third-party outstations. Five district centres are served by two telemetry computer centres using a highly resilient configuration. Provision exists within the system for the integration of local, on-site SCADA and control systems compatible with the PMCS and to a defined price offered as part of the PMCS telemetry contract. The analysis and definition of the interface to operational management systems is included as an option of the contract.

Prequalifications similar to the telemetry tender exercise took place to secure a select list of potential instrumentation and electrical contractors. Against each project 6 bidders were selected to competitively tender for the project.

The implementation programme required that a significant proportion of sites be completed prior to the telemetry system delivery and setting to work. Thus a progressive approach to site installation, testing and acceptance, communication installation and outstation installation was defined. Provision has been made for the existing telemetry and the new PMCS equipment to operate independently and, as part of the testing and commissioning strategy, sites completed early will be exercised using interim telemetry provisions.

During this extensive period working within the operational Districts, careful planning and close co-ordination between contractors and operational staff is essential. The quality of work and the safe working practices demanded require a rigorous and structured approach to site access, supervision and take-over.

CONCLUSION

The business case approach adopted by Severn Trent Water established a culture and philosophy for the PMCS project based on business need.

The establishment of the need and the benefits identified by District Managers have enabled a clear and unambiguous user requirement to be defined and agreed.

The design stage has involved a number of elements which emphasise the role of client and that of the PMCS project team. The use of the design intent documents to enable the owner, as opposed to the equipment supplier, to understand what was to be delivered has contributed significantly to the owner acceptance of the project. Combined with modern methodologies and practices has been the cultural changes in attitudes to solving the business needs rather than the exploration of the technology.

Some changes to the basic signal provision have occurred due in part to the greater emphasis on environmental and customer expectations. The business case has been flexible in that while additions have been appraised, their inclusion in the scheme is based on need and benefit.

The design activities have met the prime objective ie. that of establishing the detail necessary to describe to owners and suppliers what is technically required.

Bids are of a high quality and fall within the confidence levels expected at tendering and are within the budget.

The auditing of some 1200 sites has been successfully completed within the timescales and has generated considerable data concerning sites and the existing assets and facilities. The use of the computer databases in the business case preparation, during the design and through to the implementation has enabled greater control and flexibility in all areas of the project. Increasing levels of sophistication and applications are being generated as a result of the available information held on the computer.

The project remains on target to the original implementation programme submitted to the Severn Trent Water board in late 1988 and is within the budgets allocated.

ACKNOWLEDGEMENTS

The authors gratefully acknowledge the help they have received in the preparation of this paper from their colleagues. Acknowledgement is also made to the Directors of Severn Trent Water Ltd and Kennedy and Donkin Systems Control for their permission to publish this paper.

Chapter 11

PUMP SCHEDULING: OPTIMISING PUMPING PLANT OPERATION

T J M Moore (BHR Group Ltd, Cranfield, UK)

1 INTRODUCTION

One of the most significant costs of operating a water supply system is that of the electrical power used for pumping. The primary object of pump scheduling is to minimise this cost by utilising off-peak electricity, reducing pumping during periods when maximum demand charges are levied and by making use of efficient pumps and cheaper water sources. In addition the scheduled supply system must continue to operate within a variety of constraints; these include meeting all consumer demands and maintaining reservoir levels and mains pressures within prescribed limits. In all but the simplest cases the complexity of the scheduling problem is such that it cannot be performed effectively without a computer.

To benefit from pump scheduling a system must have the following characteristics:

- adequate reservoir storage to allow reduced pump flow during peak electricity charge periods. Perversely, although not unexpectedly, peak charge periods tend to coincide with peak water demand.

- pumping capacity which exceeds water demand so that reservoirs can be filled when required.

Fortunately the majority of water supply systems have both these characteristics.

As well as reducing pumping costs under normal operating conditions, scheduling programs introduce the possibility of decision support to aid the operator when unexpected situations arise, eg pump failure, loss of source, pipe burst, or most commonly a change in demand.

The case studies discussed in this paper have been scheduled by *PUMPLAN*, Ref.1, a scheduling software package developed at BHRGroup. *PUMPLAN* is unique, as far as the author is aware, in that it uses a KBS (Knowledge Based Systems) approach to the problem of scheduling rather than conventional mathematical optimisation methods.

2 THE PROCESS OF SCHEDULING A PUMPING SYSTEM

2.1 NETWORK ANALYSIS

Unless it is simple, with few pipes and discrete demands, the first stage in scheduling a system of any size is to build and calibrate a numerical hydraulic model, Fig.1, using a commercial network analysis package such as *WATNET*, *GINAS*, or *FLOWMASTER*. The size and complexity of the network model largely depends on what it is to be used for. For example, detailed models will be required if leakage control, pressure control, or local operational problems are to be studied.

Water Treatment – Proceedings of the 1st International Conference, pp. 105–116

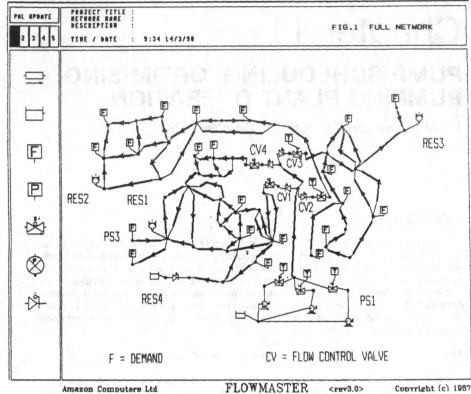

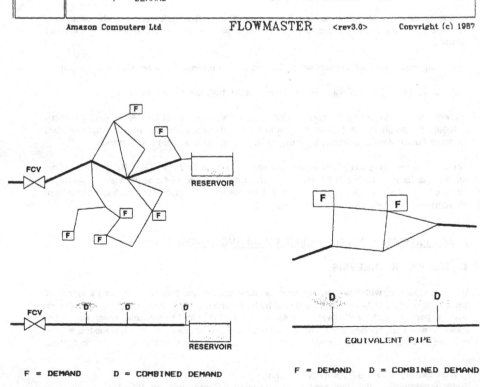

However if the network model is to be used to study pump and source optimisation, then a less detailed model which may only represent the primary trunk mains in the system is adequate.

When a full model has been satisfactorily calibrated it may be reduced to a simpler network model in various ways for use in a pump scheduling program. For example:

Demands
Where distributed demands occur, Fig.2, the full network model is used to determine what proportion of the demand is supplied at the different off-take points on the main. Comparisons should be made at several demand flows to see if the flow split between the off-takes is significantly affected.

Equivalent Pipes
Where parallel or more complex pipe connectivities occur, Fig.3, then the full network model is used to determine an equivalent single pipe.

2.2 SIMPLIFIED NETWORK TO BE SCHEDULED

The simplified network to be scheduled is shown in Fig.4; it is equivalent to the original full network shown in Fig.1. The simplification process eliminates the need for the scheduling program to undertake complex network analysis as well as schedule the pumps, whilst preserving sufficient accuracy for the scheduling to be effective; to do both would increase the processing time unacceptably.

CREATE   MODIFY   COPY   DELETE   QUIT

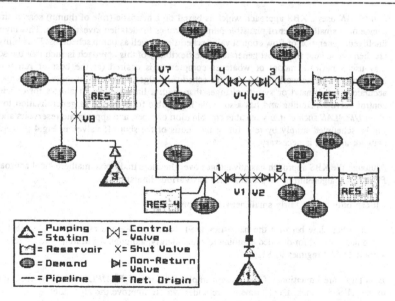

FIG.4 SIMPLIFIED NETWORK [STUDY 1]

2.3 OTHER DATA REQUIRED FOR SCHEDULING

Much of the data needed for a scheduling program will already have been collected to produce the network model. Other data required are:

- Pump performance. This has a direct effect on the pumping costs Ref.2; it is always preferable to use current site data rather than the works test data, particularly if pump deterioration is suspected.

- Electricity tariff.

- Operational constraints such as maximum and minimum reservoir levels and water abstraction limits.

3 PUMPLAN

PUMPLAN is a 286 or 386 PC based demonstrator program, produced by BHRGroup's software division SFK, which applies KBS techniques to the problem of pump scheduling. *PUMPLAN* is capable of scheduling real life systems with up to three pumping stations and five reservoirs. To date it has been supplied to two of the newly formed UK water companies.

The KBS approach has been adopted to overcome the drawbacks of traditional mathematical optimisation techniques on which most scheduling programs are based. The mathematical optimisation techniques fall into two main categories; linear programming and dynamic programming.

Linear programming methods usually converge rapidly to a solution for quite complicated problems; but the linear representation of the system hydraulic equations, ie the network model, is only an approximation. Often a much simplified bulk transfer model is generated using a network package separate from the scheduling package. This is likely to be inaccurate and will make the program inflexible to changes in demand and connectivity.

Dynamic programming is more flexible than linear programming but computing requirements increase dramatically with network complexity and this means that large networks have to be split into several smaller ones that are then scheduled separately.

PUMPLAN uses a KBS approach which is based on a heuristic (rule of thumb) search strategy to decide on a small number of plausible pump schedules for detailed investigation. This involves the intelligent generation of flow control valve schedules as well as pump schedules. Feasibility checks are then carried out and a cost generated. The flexibility of this approach is such that the scheduler may use a network model of whatever complexity is desired (at the cost of run time). The integration of the network model with the scheduling algorithm makes it more able to re-schedule accurately for demand or system changes than with a linear model. It allows for efficient flow control valve scheduling and has also made it possible for a degree of generalisation to be built into *PUMPLAN* such that any feasible combination of pipes, and appropriate reservoirs and valves, can be scheduled simply by re-defining the status of the shut-off valves in Fig.4 ; no additional network analysis is necessary.

Although the KBS approach has advantages over the more traditional mathematical approaches the *PUMPLAN* algorithm in its present form has various limitations.

- It can only handle fairly small networks at present.

- It is rather slow because the heuristics used do not reduce the number of plausible pumping schedules selected for detailed examination sufficiently; although the search space is reduced from a possible 11^{24} regimes to 81.

Both these are limitations of the present implementation of *PUMPLAN* and are not fundamental to the KBS approach. Developments are under way to improve the algorithm.

4 CASE STUDIES

4.1 STUDY 1

The network shown in Fig.1, models the extensions planned for 1995 to a medium sized water distribution system serving a population of approximately 50,000 plus several industrial demands. One of the major developments involves the construction of a new pumping station, PS1, and source. This station will supply the system with the majority of its water via a 10km long 900mm diameter trunk main and four FCVs (Flow Control Valves) V1, V2, V3 and V4. The valves are shown more clearly in the simplified network, Fig.4, used for scheduling. Three variable level reservoirs R1, R2 and R3, and a fixed level reservoir R4 are supplied from the FCVs. A second pumping station, PS3, using a fixed pumping regime, provides part of the water requirement at R1.

The network was simplified as described in Section 2.

The main objective of the study was to optimise the operation of the pumps at PS1. In addition to the pump optimisation it was also necessary to determine the opening of the FCVs such that the demand flows are met and reservoir levels controlled with the added requirement that throttling losses in the valves were to be minimised. The problem is complex even in such a relatively simple network, because the operation of any of the FCVs will affect, to a greater or lesser extent, the head and flow delivered by the pumps. As a consequence the head upstream of the FCVs and hence the flow through them will change. Solving this problem demonstrated the potential of the KBS approach; it proved possible to define rules governing the control of the FCVs such that a low loss (not always the lowest) FCV setup can be achieved iteratively. The approach allowed the effect of different rules on system operation to be examined during the development of *PUMPLAN*.

Figs.5, 6, 7 and 8 show the operation of the system under normal conditions with average day demands.

The level profiles, Fig.5, of the reservoirs R1, R2 and R3 are coincident in this case because *PUMPLAN* has intentionally been set up so that at all times there is equal shortfall or excess of water (in terms of reservoir % level) throughout the system; also the initial maximum and minimum levels of these three reservoirs are the same. The profiles show an increase in reservoir level during the off-peak electricity charge period with maximum station flow and a decrease during the peak rate and maximum demand charge periods as reserve storage is used to supplement the reduced pump flow. R4 is a constant level reservoir.

The pump schedules for PS1 and PS3 are shown in Figs.6 and 7 respectively; it can be seen that maximum advantage is being taken of off-peak electricity between 00:00 hrs and 07:00 hrs and that minimum pumping takes place for the rest of the day with the reservoirs supplying part of the demand.

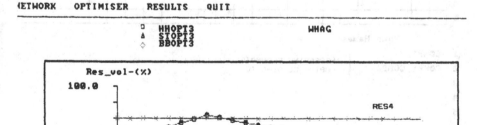

FIG.5 RESERVOIR LEVEL PROFILES [STUDY 1]

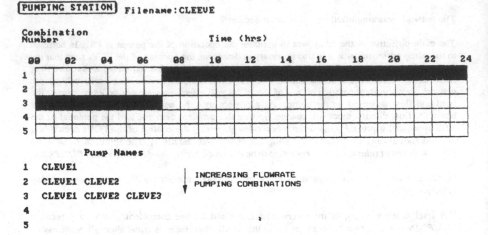

FIG.6   PS1 PUMP SCHEDULE [STUDY 1]

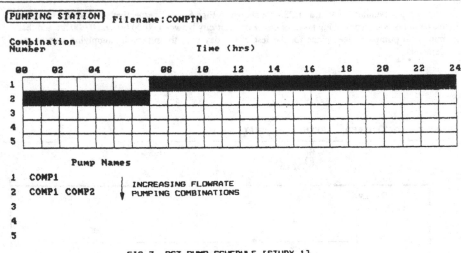

FIG.7   PS3 PUMP SCHEDULE [STUDY 1]

The corresponding valve operation schedule is shown in Fig.8. The increase in opening of valves 1, 2 and 3 at 07:00 hrs is due to a steeply peaking demand combined with a simultaneous reduction in pumping. The effect is somewhat exaggerated by the valve opening versus loss characteristics which are such that above an opening of about 0.5 the valve has a rapidly diminishing effect on the flow. After 08:00 hrs FCV4 closes completely. This is because R1 is able to meet all the demand in branch 4 and part of the demand in branch 1. The remaining demand in branch 1 is supplied via FCV1, which remains partially open.

Because this is a new system pumping costs are not available, however it is estimated that the pumping energy cost of normal operation averaged over a year will be about £90,000 if the pumps are scheduled; this is about 17% cheaper than if the system were operating under level control during the day but still retaining maximum pumping during the off-peak period.

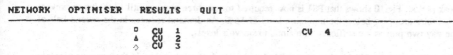

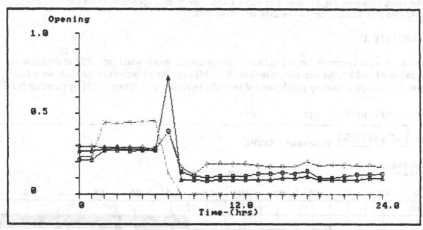

FIG.8   FCV OPENING [STUDY 1]

The use of scheduling programs in the decision support role is illustrated in Figs.9, 10 and 11 where the failure of PS3 has been simulated. Normally PS3 supplies 9.4 Ml/day of water to R1 out of a total demand of 22.4 Ml/day; PS1 has to be re-scheduled to provide an additional 72% of flow as cheaply as possible. The change in pumping will also affect the operation of the FCVs in particular those that supply R1, ie FCVs 1 and 4.

With reduced pumping capacity the reservoir level profile, Fig.9, is now very much flatter and even by running the maximum of three pumps PS1 cannot significantly increase levels during the off-

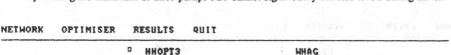

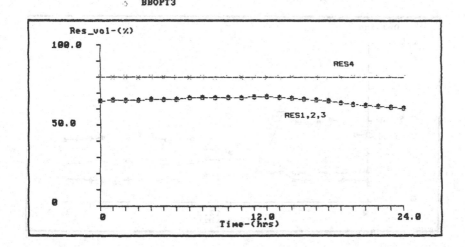

FIG.9   RESERVOIR LEVEL PROFILES [STUDY 1]

peak period. Fig.10 shows that PS1 is now required to run three pumps until 13:00 hrs to prevent reservoirs dropping below the minimum level of 55%. As demand reduces in the second half of the day two pumps are sufficient to maintain reservoir levels.

The valve openings, Fig.11, show that FCVs 1 and 4 are further open than in the previous example to enable PS1 to make up for the flow loss from PS3.

4.2 STUDY 2

The trunk main system shown in Fig.12 is currently used to pump water from PS1 to the treatment works at R1 and R5; the total system demand is 55 Ml/day. The objectives of the study were firstly to reduce pumping costs by scheduling and secondly to reduce abstraction at PS1 by pumping from

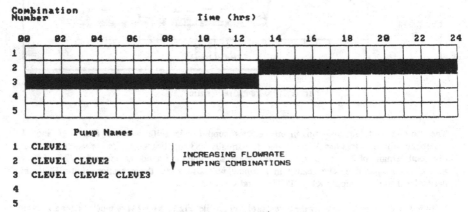

FIG.10 PS1 PUMP SCHEDULE [STUDY 1]

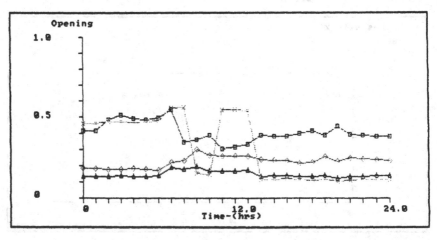

FIG.11 FCV OPENING [STUDY 1]

CREATE   MODIFY   COPY   DELETE   QUIT

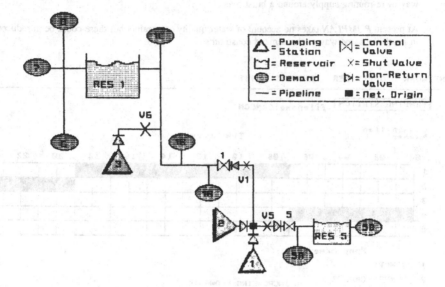

FIG.12   SIMPLIFIED NETWORK [STUDY 2]

PS3. This station had been little used for a number of years; integrating its operation with that of PS1, coupled with recent pipe modifications at R1, considerably increased the scope for pump scheduling in this system. Under certain circumstances the water quality of one of the sources at PS1 limited the useable pump combinations available at this station.

PUMPLAN showed that by capitalising on the increased pumping capacity now that PS3 was available and by scheduling both stations, a 14% reduction in the current annual pumping energy cost of over £400,000 could be achieved. At the same time abstraction rates at PS1 were reduced to a satisfactory level.

The schedules for PS1 and PS2, Figs.13 and 14, show that reduced pumping has been achieved between 07:00 hrs and 19:00 hrs when both peak rates and maximum demand charges apply. Some increase in pumping is necessary between 19:00 and 24:00 hrs to prevent reservoir levels dropping below the desired start level for the following day; although peak rate electricity is being used, the effect on pumping cost is not as serious as it would have been had the additional pump been switched on during the maximum demand period. The corresponding reservoir profiles are shown in Fig.15 and reflect the changes in pumping combination throughout the day.

5 PUMP SCHEDULING - THE FUTURE

The privatisation of the water industry and new legislation are creating a need for continued improvement in the efficiency and reliability of water supply system operation and also in water quality. These improvements will necessitate more sophisticated control of the day to day operation of supply systems; it is in the operational decision support role that scheduling programs will become increasingly important.

Decision support programs will not only be required to give the operator advice on optimising normal operation but also in unforseen circumstances. Changes in demand and pump status, for instance, are normally easy for scheduling programs to deal with, however pipe bursts can present much greater difficulties. This is an important area where the KBS approach is expected to have major advantages over the traditional programming methods, because it has the potential to be able to handle unplanned changes in connectivity. A further possibility is that of advising on the best way of re-routing supply around a burst main.

At present *PUMPLAN* takes no account of water quality constraints but these could be introduced in much the same way as the demand constraints.

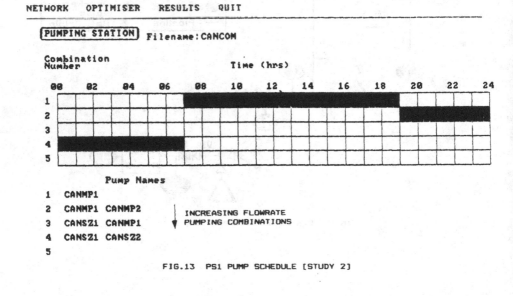

FIG.13 PS1 PUMP SCHEDULE [STUDY 2]

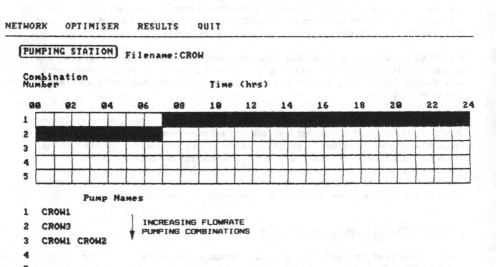

FIG.14 PS3 PUMP SCHEDULE [STUDY 2]

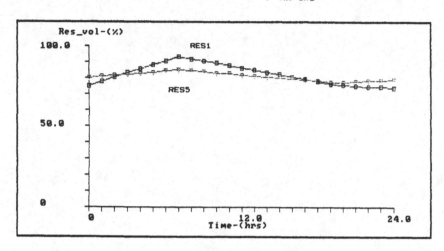

FIG.15   RESERVOIR LEVEL PROFILES [STUDY 2]

6 CONCLUSIONS

Although primarily intended to demonstrate BHRGroup's innovative approach to pump scheduling, *PUMPLAN* has already proved itself as a useful tool in commercial applications.

The KBS approach allows the possibility of extremely advanced decision support software to be considered that will be capable of advising on a much wider range of unplanned situations than had previously been thought feasible.

The philosophy that it is more important to find an operationally good schedule that has been based on rigorous network analysis than to find a mathematical optimum, based on a less accurate network model, helps to overcome some of the drawbacks inherent in the software presently available.

REFERENCES

1] M.J.Likeman, T.J.M.Moore "A Knowledge Based Approach to Pump Scheduling" BHRGroup 1990.

2] T.J.M.Moore, P.N.Smith "Methods of Assessing and Reducing the Cost of Water Supply Pumping" BHRGroup Report N° RR3207 Jan 1990.

Chapter 12

GAS LIQUID MIXING AND MASS TRANSFER IN THE WATER PROCESSING AND CHEMICAL INDUSTRIES: A BRIEF COMPARISON OF THE TECHNOLOGIES

A J Green and M J Whitton (BHR Group Ltd, Cranfield, UK)

1. ### Introduction

The aim of this paper is to briefly review, compare and contrast some of the technologies employed in the aerobic treatment of wastewater, against those used by the chemical/pharmaceutical industries for gas liquid mass transfer. The paper does not profess to provide a detailed analysis of the pros and cons of the technologies, but more suggest areas where transfer from one industry to another may be of value.

The paper limits its scope to devices employing gas in liquid dispersions, rather than liquid in gas dispersions or liquid films. Thus such devices as packed columns, plate columns, trickling filters etc. are not covered.

2. ### Gas Liquid Mass Transfer

The rate of mass transfer of a specified component, e.g. oxygen, from a gas to a liquid can be described by:-

$$J = K_L a V_L (C_L^* - C_L) \qquad (1)$$

where

J = mass transfer rate (kg s^{-1})

K_L = liquid side mass transfer coefficient (m/s)

V_L = volume of dispersion (m^3)

a = interfacial area/unit volume (m^{-1})

C_L^* = Interfacial concentration of the specified component in the liquid (kg m^{-3})

C_L = concentration of the specified component in the bulk of the liquid (kg/m^3)

Water Treatment – Proceedings of the 1st International Conference, pp. 117–128

For the aeration of water, the mass transfer rate must be sufficient to satisfy the oxygen demand, both chemical and biological, over the required treatment period. Equation 1 identifies the main controlling variables in the process:-

1. The mass transfer coefficient (K_L). This is affected to some degree by the level of turbulence in the liquid, but more significantly by the liquid physical and chemical properties. Thus K_L will depend on the effluent composition.

2. Interfacial area (a). This will depend on the degree of turbulence in the system and the fluid properties. The physical and chemical properties of the liquid medium influence bubble coalescence behaviour and hence bubble size distribution.

3. Concentration driving force ($C_L^* - C_L$). For the calculation of $k_l a$ the interfacial concentration is assumed to be the liquid concentration which is in equilibrium with the bulk gas concentration. The concentration driving force can be increased either by increasing the percentage of oxygen in the gas (eg use of pure oxygen rather than air) or by pressurising the gas (eg by increasing tank depth).

To cope with the uncertainties of 1 and 2, many manufacturers quote a so called "α-factor", which is the ratio of oxygen transfer rate in the actual effluent to that in clean water (where most development work has been carried out). K_L and a are difficult to measure independently so the vast majority of literature data are reported as the product $k_l a$.

In addition to the primary purpose of supplying oxygen, mixing systems often have to satisfy a number of additional functions.

(i)   Good liquid mixing to prevent 'dead zones'.

(ii)   Suspend solids.

(iii)   Strip out Carbon Dioxide produced by the bacteria.

3.   **The Differences between Water and Chemical Process Requirements**
Requirements

Before looking at the techniques employed in the two industries, it is worth identifying some significant differences between the requirements. These, in general terms, are as follows:-

1. The value of the product. Water is probably the ultimate "low value" product. As such, running and capital cost are the key considerations in the design. For higher value chemical, and particularly pharmaceutical, products cost is of lower importance. If a system can be designed, for example, to increase product yield, this is very likely to overwhelm any increase in development or capital/running costs.

2.   The scale of operation. The size of operation of waste treatment equipment may be orders of magnitude greater than a chemical production plant, although the largest vessels may be a few 10's of metres in diameter. Direct scaleup of a chemical/pharmaceutical plant geometry to that required for water treatment may not be feasible from a physical/mechanical engineering standpoint. Additionally, at large scale, water depth considerations become very important, because of the increased pressure required from any air supply to cope with the head of water.

3.   The rate of mass transfer and intensity of mixing employed. For water treatment, only slow mass transfer is required to supply the biological demands. Typical values of $K_L a$ are of the order of 10^{-3} s^{-1}. For chemical/pharmaceutical operations, values up to 0.1 s^{-1} are common.

These differences will be considered further in the Discussion Section.

4.   Chemical and Pharmaceutical Techniques

4.1 Agitated Vessels

The "workhorse" of the chemical process industry for gas liquid mass transfer is the agitated mixing vessel. Although a wide range of geometries are used, two of the most common are illustrated in Figures 1 and 2. Tall vessels, eg Figure 2, have a greater surface area/unit volume than the standard geometry vessels, Figure 1, and are therefore used when heat transfer is critical, as is the case in some pharmaceutical processes.

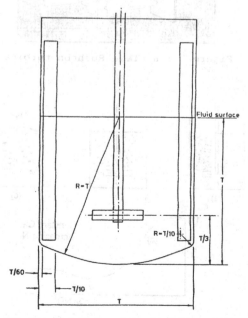

Figure 1 : Standard geometry vessel with a single impeller

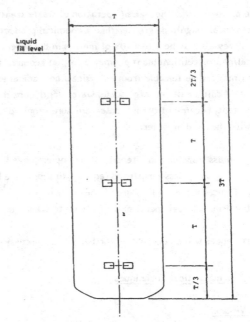

Figure 2 : Tall vessel fitted with multiple impellers

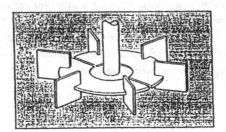

Figure 3 : 6 blade Rushton turbine

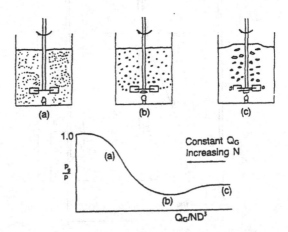

Figure 4 : Gas liquid flow patterns

The most commonly used impeller for gas liquid applications is the Rushton turbine, Figure 3, which has a mostly radial flow pattern . Such systems suffer a number of limitations:

(a)   As gas rate increases for a fixed impeller speed the power drawn by the impeller decreases, Figure 4. Its ability to disperse gas reduces, Figure 4 (b), until it becomes overwhelmed by gas or flooded, Figure 4 (c).

(b)   It produces a relatively small amount of liquid flow compared to its power.

(c)   In multiple impeller configurations, there is a tendency for flow to compartmentalise, resulting in poor top to bottom mixing.

As a result of the above, there is currently a great deal of interest in developing alternatives to the Rushton, including the scaba SRGT, Figure 5 (a), (said to keep constant power draw with increasing gas rate, Ref 1) and axial flow turbines such as the Prochem, Figure 5 (b) or the Lightnin A315. Work is underway at a number of institutions, notably Fluid Mixing Processes (FMP) at BHR Group, to improve design data for existing and novel arrangements.

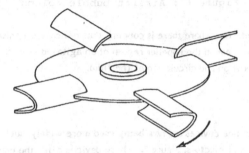

Figure 5(a) : Scaba 4SRGT turbine

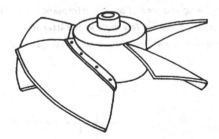

Figure 5(b) : Prochem hydrofoil

4.2 Airlift Reactors

An alternative to the agitated vessel is the airlift reactor illustrated in Figure 6. In such a design the upflow of air drives the fluid around a loop, either internally within the vessel, or externally. Such systems are widely used in the

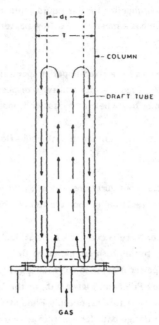

Figure 6 : Airlift bubble column

biotechnology industry, where there is concern that delicate organisms will be
damaged by fluid shear in the impeller region of an agitated vessel. The design
provides good aeration and circulation of the fluid.

4.3 In Line Devices

High intensity, in line devices are now being used more widely, such as
motionless mixers or ejectors, Figure 7. These devices offer the opportunity
to increase processing rate and/or reduce plant scale ("Proces
Intensification"). Scale up and prediction of performance of such devices is
claimed to be simple, but there are few data in the literature. As a result,
work has commenced recently at BHR Group to investigate such devices in
more detail, Ref 2.

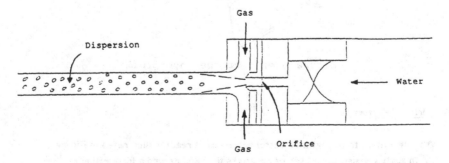

Figure 7 : Gas liquid jet ejector

5.   **Waste Water Processing Techniques**

5.1 Submerged Turbine Aerators

These are the water industries equivalent of the agitated mixing vessel.
Various designs are on the market, ranging from axial to radial flow turbines,
with an air sparge beneath, usually aeration units are mounted within square or
rectangular tanks, often with multiple units. Often the units are fitted with
some form of baffling.

The turbine performs multiple roles:

-       breakup and dispersion of sparged air

-       mixing of liquid in the tank

-       suspension of sludge, preventing settlement

One disadvantage of such systems is that additional equipment (i.e. a blower) is
needed to introduce the air. To overcome this, various designs of self-
aspirating turbines have been developed, e.g. Ref 3. However, these are
generally less efficient at pumping the liquid.

5.2 Surface Aerators

These devices are similar to submerged turbines, but are positioned at or near
the surface. Air is drawn down from atmosphere and pumped into the tank by
the agitator, Figure 8 (a). These can either be mounted on a fixed frame or
float on the surface. They have the advantage of lower capital and running
cost than submerged turbines, but require more freeboard, due to the extensive
surface disturbances caused, and (if on a fixed frame) are less able to cope
with fluctuating liquid levels. Additionally, the devices have less influence on
the base of tanks, so suspension performance is reduced.

As an alternative, "brush" aerators have been used, Figure 8 (b), i.e. impeller
shaft mounted horizontally. In these designs mass transfer occurs through both
liquid in air and air in liquid dispersion. They are often used in oxidation
ditches, where a linear water flow is required, this being enhanced by the
aerator.

Finally plunging jet devices, Figure 8 (c), have been found to have great
potential, Ref 4, but have yet to find their full commercial potential.

5.3 Bubble Diffusers

These can be separated into two categories: coarse bubble diffusers,
comprising arrays of perforated sparge pipes, and fine bubble diffusers,

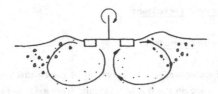

Figure 8(a) : Mechanical surface aerator

Figure 8(b) : Brush type surface aerator, ref. (11)

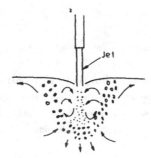

Figure 8(c) : Plunging jet aerator

comprising fine-pore, large open area porous materials (e.g. ceramics, plastics, sintered metals). The latter are often designed to cover the entire tank floor, to ensure even dispersion of air within the tank and avoid sludge settlement, Ref 5. This removes the need for good liquid phase mixing. Reference 10 recommends tanks with large length to width ratios, resulting in an approximation to plug flow (i.e. all effluent has the same residence time). Conversely, effluent mixed in a large square tank will have a very wide range of residence times.

5.4 Other Techniques

There is a wide range of other techniques currently used, including the following:-

(a) Ejectors. A technique has been developed by the Electricity Council Research Centre which employs an ejector to achieve rapid mass transfer, the jet then being used to mix the tank.

(b) Fixed Head Aerator. This device was developed by BHRA, Ref. 6, and uses similar principles to (a), i.e. achieving rapid, localised mass transfer, then mixing the tank with the jet, Figure 9.

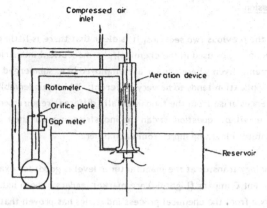

Figure 9 : Arrangement of fixed head aerator, ref. (6)

(c)   ICI "deep shaft" process. This operates in a similar manner to the airlift
       reactor, Section 4.2, see Figure 10. Very large depths are employed (up
       to 100m), giving very high partial pressures in the gas phase, and hence
       rapid mass transfer.

(d)   Enriched oxygen techniques. Enriched air, or pure oxygen, are used to
       increase the mass transfer rate by increasing the concentration during
       force (e.g. BOC Vitox process or the Unox process).

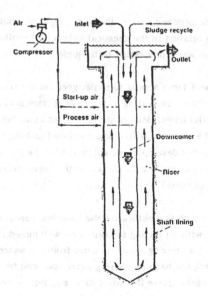

Figure 10 : Deep shaft aerator, ref. (12)

6.   Discussion

From the previous two sections, it is clear that there is little commonality between
the technologies used in the chemical/pharmaceutical industries and waste water
treatment. Even where similar are employed (e.g. submerged turbine aerators),
their application tends to be very different. Whilst, undeniably, many of the
differences arise from the fundamentally different requirements outlined in Section
3, the inevitable question is: can the industries improve their processes (and
profitability) learning more from each other?

Technology transfer at the manufacturer level is good: for example, Mixing
Equipment Company (Lightnin) are market leaders in both industries. However,
evidence from the chemical process industries has proven that there is great
benefit in the equipment end users themselves improving their knowledge of
processes: the success of process industry related research club projects, such as
FMP and HILINE at BHR Group and others (e.g. HTFS, SPS etc) is a good example.

Looking at two recent conferences related to mixing provides some
enlightenment. A conference of "Bioreactor Fluid Dynamics", Ref. 7, attracted no
papers and no delegates from the water industry; yet what is an activated sludge
treatment plant, if not a bioreactor? The most recent European Conference on
Mixing, Ref. 8, attracted only 2 papers as waste water applications out of a total of
77. Equally, recent conferences related to wastwater treatment probably attracted
few delegates from the chemical industry. Process industry research clubs have
had little success in recruiting members from the water industry, although FMP and
HILINE have recently both recruited members (Yorkshire Water and C.G.E.
respectively).

A detailed analysis of the pros and cons of the technologies will not be tackled, but,
speaking from expertise relating to the chemical industry, the following
observations are made on water processing techniques:-

1.   Many water treatment tanks appear to be designed totally around civil
      engineering considerations, resulting in square or rectangular, relatively
      shallow tanks with flat bases. Mixers are introduced as an "add on extra".
      Could more effort be placed in designing a combined tank/agitator system,
      closer to the geometries developed and optimised for the chemical industry?
      This may result in higher civil costs, but would the improved/reduced
      processing cost compensate?

2.   Aeration techniques such as ejectors and the fixed head aerator employ local,
      rapid mass transfer with jet mixing to keep tanks well mixed. The original
      technology for liquid jet mixing actually comes from the water industry
      (mixing in reservoirs), but has since been greatly advanced for applications
      such as oil storage tanks. Have the latest data, e.g. Ref. 9, been transferred
      back to the water industry to aid designers?

In summary, to quote reference 10, could it be that

"..... mechanical devices, economy of thought or love of concrete have ruled design".

7.   Concluding Remarks

As stated at the outset, this paper has not intended to provide deep analysis and insights into the differences between chemical pharmaceutical processes and waste water treatment.

However, it is hoped that, by setting down the different techniques used in both industries, it will prompt members of both, and particularly the water industry, to think about how they might use each other's expertise and investigate the literature and other sources of data more fully. If so, the paper will have achieved its purpose.

References

1.   "Gas Dispersion Performance in Fermenter Operation"A.W. Nienow, Chem. Eng. Progress, February 1990.

2.   "High Intensity In-Line Mass Transfer Equipment, A Collaborative R&D Programme: HILINE. Proposal to Potential Members". A.J. Green, BHR Group, June 1988.

3.   "Sorption Characteristics for gas-liquid contacting in mixed vessels" M. Zlokarnik, Adv. in Biochem Eng. $\underline{8}$ 134 (1978)

4.   "The effect of contaminants on the oxygen transfer rate achieved with a plunging jet contactor" J.A.C. van de Donk, R.G.J.M. Lans, J.M. Smith, 3rd European Conference on Mixing, York, BHRA, pp 289-297 (1979).

5.   "Energy Saving by Fine-bubble Aeration" B. Chambers & G.L. Jones. Water Pollution Control (1985).

6.   "The Development & site testing of an aerator", R. King & H.C. Young. Symposium on the Profitable Aeration of Waste Water, London. BHRA pp 31-54, April 1980.

7.   "Proceedings of 2nd International Conference on Bioreactor Fluid Dynamics" R. King (ed), Cambridge. BHRA 1988.

8.   "Proceedings of the 6th European Conference on Mixing", Pavia, Italy, 1988. Distributed by BHRA.

9.   "Jet Mixing Design Guide" R.K. Grenville, BHRA (confidential to FMP members).

10.   "Putting Chemical Engineering into Aerobic Effluent Treatment -2" R. Olley. The Chemical Engineer, July 1989.

11.   "The aeration of waste water: Basic physical functions". A.W. Nienow Symposium on the profitable aeration of waste water, London, BHRA, April 1980, pp 1-12.

12.   "Fermentation reactors", W. Sittig, Chemtech, Oct. 1983, pp 606-613

PART IV

Control and measurement techniques

Chapter 13

OPTIMISATION OF SECONDARY CHLORINATION PRACTICE IN THE UK

C R Hayes (Water Quality Management, UK), A Elphick (Consultant, UK)

SUMMARY

The need for boosting chlorine concentrations within water supply distribution networks is reviewed in relation to UK practice and recent changes in the UK's water legislation. Comparisons are made between UK practices and those of other European States.

There is a need to clarify better several microbiological quality issues, in order to be able to define more objective approaches to the management of chlorination. Irrespective of these current uncertainties, it is concluded that secondary chlorination will be increasingly necessary in the UK, at least in the short to medium term. This view relates to the technical challenge of achieving microbiological, aesthetic (eg. taste) and chemical (eg. trihalomethanes and nitrite) standards within the next few years to satisfy legislative demands.

An objective approach is proposed for managing chlorine residuals within distribution networks, based on the specification of target minimum concentrations. Its implementation in any network will require preliminary investigations to determine the need for secondary dosing; if such dosing is required, it will then be necessary to select appropriate doses and dosing points. Such investigations might be supported by a secondary chlorination model.

The design and use of chlorination plant, in respect of secondary dosing applications and for optimisation studies, are reviewed.

INTRODUCTION

The benefits of maintaining chlorine residuals throughout a water supply distribution network have often been perceived to be :

(i)   the suppression of bacterial regrowth and the preservation of general sanitary condition

(ii)   the provision of a buffer against low level contamination

(iii)   the presence of a chemical indicator to detect gross contamination

Water Treatment – Proceedings of the 1st International Conference, pp. 131–140

These perceived benefits have been examined in conjunction with
the legislative standards that apply to microbiological quality,
in order to establish the objectives for disinfection processes.
In this examination, comparisons have been made to other European
practices.

The constraints of other quality standards, such as taste,
nitrite and trihalomethanes, have also been considered to help
determine the most likely means of achieving disinfection goals.
From this review, it is evident that secondary chlorination will
be increasingly necessary in the short to medium term.

Optimisation of secondary chlorination will require the adoption
of a more objective approach to managing disinfection residuals
within distribution networks and preliminary field study,
techniques and plant for which are discussed. The scope for
developing an optimisation model is identified.

LEGISLATIVE   ASPECTS   OF   DISINFECTION

Disinfection Practice

In the UK, it has been considered necessary that all public water
supplies should be disinfected, national guidelines (DoE,1983 and
WAA,1988) reflecting this view. The recent Water Supply (Water
Quality) Regulations 1989 maintain this approach as a legal
requirement, although the Government can authorise for specified
groundwaters not to be disinfected. The national guidelines have
also drawn attention to the need (DoE,1983) or advantage (WAA,
1988) of maintaining disinfectant residuals throughout
a distribution network.

In establishing these general requirements for disinfection,
neither the national guidelines nor the Regulations, indicate the
type of disinfectant to be employed, its dose and contact time
or the concentration of the residuals to be maintained. Rather,
it is left to the individual water undertakers to manage
disinfection such that microbiological standards are achieved.

The integrity of microbiological standards is closely related to
sampling frequency, which has varied across the UK due to
ambiguous sampling guidance (DoE,1983). In consequence,
disinfection practice has varied also. To compound matters,
several of the microbiological parameters are poorly defined.

The most explicit guidance for managing disinfection is provided
by the World Health Organisation (WHO,1984). For primary
disinfection at sourceworks it is recommended that between 0.2
and 0.5 mg/l free chlorine (the higher concentration is desirable
for "unprotected" sources) is maintained for a minimum 30
minutes contact period with turbidity <1NTU and pH <8.0.

Generally, these requirements are achieved in the disinfection
of surface derived waters in the UK although the turbidity and
pH requirements are probably not achieved consistently at some
works. Also, many groundwaters are disinfected without a contact
period prior to entering distribution.

The WHO also recommend the maintenance of a free chlorine
residual of 0.2 to 0.5 mg/l in the distribution system in order
to reduce the risk of microbial regrowth and to provide an
indication of the absence of post-treatment contamination; this
recommendation is commonly not achieved in UK distribution
networks, whether by design or default.

In order to achieve greater consistency and objectivity in the
management of disinfection, including secondary chlorination,

consideration must be given to the water quality objectives of such treatment; these are discussed below.

Microbiological Standards

The EC "drinking water" directive (EC,1980) requires all samples of water at the point of use to be free of faecal coliforms (E.coli) and pathogens, and that at least 95% of such samples to be free of total coliforms. The recent UK Regulations implement these standards although routine monitoring for pathogens is not specified.

The coliform indicator approach to microbiological quality control has been generally successful in minimising water borne disease in the UK (Galbraith et al,1987). However, possible limitations of the approach are increasingly being realised : eg with respect to entero-viruses (WRc,1986) and protozoan parasites (DoE,1989). The main weakness, however, in relying solely on coliform standards for determining disinfection practice, is the uncertainty that is inherent in any water sampling.

The confidence to be expected in a given sampling frequency can be estimated statistically using the properties of the binomial distribution (A E Warn, personal communication); such estimates indicate that there is a risk of poor water quality failing to be noticed at low sampling frequencies, as illustrated in Table 1.

Table 1.   Statistical estimates of true failure rates giving zero failed samples as a function of sampling frequency.

Number of samples	Most likely true failure rate (%)
6	10.9
12	5.6
26	2.6
52	1.3
300	0.2

It must thus be realised that the minimum sampling frequencies specified by the EC directive and the UK Regulations carry risks of failing to notice poor quality, particularly for smaller supplies for which the specified minimum frequencies are lowest. The authors conclude that, in seeking to achieve high standards of microbiological quality, additional approaches must also be followed.

The EC directive contains guide levels for 22 and 37 C bacterial colony counts. These have been virtually ignored in the UK and the new Regulations only require "no significant increase over that normally observed "; this approach may have some validity to samples taken from a fixed point but is difficult to apply to random samples, particularly those from surface-water derived supplies in which the potential for bacterial regrowth is often high. Greater recognition of the possible significance of colony counts could be warranted in the UK for the following reasons:

(i)   high colony counts may be indicative of the presence of spoilage organisms of concern to hospitals and the food and drink manufacturing industries

(ii)   high colony counts may be indicative of the presence of Aeromonas hydrophilla, being of possible health concern (Burke et al,1984)

(iii) high colony counts may be indicative of conditions suitable for infestation by aquatic animals (eg Nais worms)

Conditions that lead to elevated colony counts include the presence of assimilible organic carbon (AOC), the stagnation of water and the attenuation of effective chlorine residuals which may be exacerbated in extensive distribution networks.

Such deterioration has been particularly appreciated in the Netherlands where AOC control is practised at several works; by reducing the driving force for bacterial growth, it has been possible to substantially reduce the need for carrying disinfectant residuals throughout distribution, the ultimate example being the cessation of "safety" chlorination for supplies to Amsterdam (Schellart,1987). It is also worth noting that in the Netherlands, tentative standards for Aeromonas hydrophilla have been promulgated due to health concerns (A Hulsman - personal communication).

Of further relevance to disinfection practice is the requirement of the EC directive for pathogens to be absent. Whilst this requirement is also contained in the UK's new Regulations, neither item of legislation defines routine monitoring clearly, albeit the EC directive does state that such monitoring, in a supplementary manner to coliforms, may be necessary. Recognising that some pathogenic micro-organisms could be present in the absence of coliforms, it would seem appropriate for further clarification of minimum sampling requirements to be provided by the regulatory bodies; such a development could materially influence disinfection practices.

Aesthetic and Chemical Constraints

Despite general recognition that microbiological quality is paramount, and therefore presumeably the means of its achievement, an antagonistic range of aesthetic and chemical standards make the fulfillment of microbiological objectives potentially difficult. Such constraints are :

(i)   the desirability of minimising chlorination by-products, various standards for trihalomethanes having been applied throughout Europe; the operational implication is to reduce the use of chlorine with the potential for poorer microbiological quality

(ii)   the standard for nitrite of 0.1 mg/l NO_2 which, because of its implementation as an absolute maximum,probably precludes the use of chloramine in many supplies due to nitrite formation associated with the ammonia used and/or the breakdown of chloramine - this constraint appears, in the UK, to be the cause of a move away from chloramine to free chlorine as a disinfection residual in distribution, with the identified need for many more secondary chlorination installations (Schroders,1989)

(iii) the standard for taste of a dilution number of 3 is presently being applied in the UK to dechlorinated samples (DoE,1989b); as such the standard should have little constraining effect on the management of free chlorine residuals within distribution - however, the use of dechlorinated samples may not be consistent with the EC directive and any future change to unaltered samples could impact on chlorine residual management; the practical implications would then be the need to maintain effective residuals by applying smaller chlorine doses at more points (ie secondary chlorination).

(iv)   the EC directive requires the absence of "animacules and algas" although the UK's new Regulations offer no clear standards. The UK Government has stated (DoE,1984) that water undertakers should "take steps to control their presence"; such steps have often included the dosing of pyrethroid insecticides but implementation of the EC directive's pesticide parameter could preclude such dosing

- by implication, higher concentrations of disinfectant residual throughout distribution can be anticipated and therefore more secondary chlorination installations.

THE SCOPE FOR RATIONALISING THE MANAGEMENT OF CHLORINE RESIDUALS IN DISTRIBUTION

Perceived Operational Benefits

Microbiological quality control is essentially restrospective due to the time taken to perform microbiological tests, a review of data over a period of time helping water undertakings to determine the integrity of their operational procedures. In such judgements, an assessment of the risks of microbial ingress must be included together with the operational measures that can provide some form of protection. Undoubtably, good design and maintenance of supply networks are a prerequisite to consistently high standards of microbiological quality. Additionally, the maintenance of an "effective" disinfectant residual throughout distribution is perceived to offer protection against low level ingress as well as avoiding regrowth problems. Assuming these objectives to be valid, it is reasonable to conclude that minimum disinfectant residual concentrations can be specified as targets to suit the possible disinfectant types as well as for different water types (primarily surface or ground derived). Such an approach is less sensitive to the other major perceived benefit of maintaining disinfectant residuals in distribution; namely, the indication of gross contamination through total loss of the residual.

Chlorine Concentration Targets for Water in Distribution

In order to define numerical targets, the biocidal effectiveness of the different forms of disinfectant residual must first be appreciated; a simple classification is offered in Table 2.

Table 2   Biocidal Effectiveness of Disinfectant Residuals

Type	Biocidal Effectiveness
Free chlorine	Good
Monochloramine	Moderate
Organochloramine	Poor
Chlorine dioxide *	Good
Chlorite	Poor
Ozone **	Good

* prone to reversion to chlorite
** unstable

On the basis of this classification, whilst free chlorine, chlorine dioxide and ozone all offer sound options for primary disinfection of source waters, only free chlorine and monochloramine have the stability and biocidal effectiveness to be considered as residuals for distribution network protection. Chlorine dioxide is used at several European waterworks as a "safety" disinfectant but the authors question this practice for the reasons stated above.

Experience in Europe is increasingly demonstrating the benefits of AOC removal in treatment and the subsequent reduced requirement for safety disinfectant residuals to be maintained. In several European States, groundwaters are not, generally, disinfected at all, raising fundamental questions about the need for maintaining disinfectant residuals in distribution. In comparing these countries to the UK, it may be the case that the more highly developed approaches to protecting raw water resources through protection zones, as in the Netherlands and West Germany, permit the "need" for safety disinfection to be avoided; further, there may be differences in the general condition of distribution networks. However, for the time being, the maintenance of disinfectant residuals in UK systems is anticipated.

With regard to the above arguments, the authors propose (in Table 3) a set of chlorine and monochlorine target concentrations which can be applied to all distribution systems; how the targets are achieved will be dependent on the individual circumstances of each distribution system.

Table 3 Proposed Target Minimum Concentrations for Free
        Chlorine and Monochloramine

Water type	Minimum free chlorine residual* mg/l	Minimum mono-chloramine residual* mg/l
Ground	0.05	0.15
Surface	0.10	0.30

* to be achieved for at least 95% of the receiving population

Given that AOC control was effective for a surface-derived water, then the ground water targets could apply. Given that catchment protection measures were in force and that the distribution system was in sound condition then these targets could become unnecessary, ie the practices of other European States could follow.

Maximum concentrations of 0.5 mg/l free chlorine and 1.5 mg/l monochloramine could be debated; these will depend largely on the skill of the public relations department (taste complaints) and cost.

It should be stressed that the targets proposed in Table 3 are based on the general operational experience of the authors; validation of such targets will require input from microbiological specialists and a consensus view on risk.

OPTIMISING SECONDARY CHLORINATION

The Need

As discussed above, compliance with microbiological standards, other microbiological issues and the constraints of aesthetic and chemical standards have combined in the UK such that the boosting of disinfectant residuals in distribution is likely to be more widely practiced. Of the disinfectant options available, greater emphasis will continue to be directed towards the use of free chlorine.

Facilitating this need can only be achieved in a consistent manner, thereby normalising the risks posed by ingress and regrowth, if an objective approach is adopted (as proposed above) throughout the UK. This implies considerable evolution of the currently vague national guidelines.

Investigations

Designing a cost-effective secondary chlorination strategy for a

given distribution network will require preliminary studies
in order to ensure that the correct chlorine doses are applied at
the correct locations. Factors that need to be considered are the
times of travel of water in the network, the internal condition
of pipes, both of which affect the rate of attenuation of
chlorine. Such investigations are likely to require a mobile
chlorine injection system, options for which are discussed below.
As temperature will influence attenuation, studies must encompass
the range in seasonal conditions.

The extent of investigation will depend on the size and
complexity of the distribution network and may be fairly onerous.
To help minimise such work, a chlorine attenuation model could
be developed which linked network characteristics and condition
to chlorine attenuation rates and thereby worked out required
dosing points; in many cases, preferred points of dosing will
not coincide with convenient structures such as service
reservoirs.

EQUIPMENT FOR SECONDARY CHLORINATION

Even where dosing is possible at a service reservoir, the
accurate application and control of chlorination may be
difficult. The addition of chlorine to a large body of water,
such as a service reservoir, to ensure good mixing in an
existing structure, may be very difficult where the inflow and
outflow vary from zero to maximum. Equally, the application of
chlorine to reservoir outlets (and there may be several) may be
very difficult for physical reasons or, again, for reasons
of large flow variations.

Dosing points need to be chosen with care. The location of dosing
points should avoid the need to install equipment in areas
accessible to the public, although secure kiosk or cubicle
arrangements have been successfully used.

Some of the requirements for adequate primary disinfection, in
particular, contact time, can be avoided in secondary
chlorination. This enables the dosing/mixing/analysis/alarm data
transfer to be more easily applied, being essential features of a
well designed secondary chlorination system.

Clearly the type of equipment used will be dependent on size and
location but in all cases it needs to be simple, reliable, easy
to install and maintain and as self-contained as possible.

Preliminary investigations with mobile plant are likely to be
unavoidable unless blanket treatment can be applied successfully
to all service reservoirs/networks.

Typically, a mobile chlorination unit is supplied as a trailer,
although portable cabins have also been used. Such units need to
be secure and stable and four wheeled trailers fitted out with
chlorine storage system, gas or chemical dosing units,
generator/booster pump set, fume detection systems etc., are in
use.

A variety of chlorinators and injectors, with boosted water
available at various pressures and volumes will allow for a wide
range of treatment conditions into points of application ranging
from zero back pressure to high pressure mains. Such application
can pin-point the most desirable dosing point if it is carried
out in conjunction with the analysis of residuals in the network
at different times.

Equipment for the chlorine residual analysis can be housed in the
mobile unit or in discreet sensing stations at relevant points in
the network.

Mobile equipment of this complexity will not only be of value in secondary chlorination assessment but, for example, if de-chlorination equipment is included, could also be satisfactorily used for mains disinfection. Similarly it could be made available for use in emergencies where rapid measures for disinfection are required.

Panel mounted, or skid, arrangements to provide both handling, dilution and dosing of hypochlorite, as well as the analysis and control of residuals in the network, are becoming subject to more exhaustive specification by water companies in the UK. Often specified is the requirement for instantaneous alarm should the chlorine residuals stray from the optimum. In all such systems accuracy and reliability is paramount and UK specifications are now likely to fall in line with the Water Authorities Association Process Systems Committee requirements for process control and monitoring instruments. These strict requirements could be further enhanced by triple validation of primary sensing signals at sensitive or problematical locations.

Secondary chlorination generally calls for the use of simple controllers (for cost and reliability). The use of sophisticated computer controlled systems in remote locations has its advocates but the relative importance of re-chlorination compared to primary disinfection must be borne in mind.

On site electrolytic chlorination (OSEC) is likely to reduce the need for visits to these secondary chlorination sites. Self-contained, easily installed systems are now available at a wide range of capacities down to 2 kg per day. Modern technology allows economic conversion of electrical energy and salt into hypochlorite with fully integrated control and alarm systems for local and remote monitoring. Storage of salt, instead of hypochlorite, provides stability of chemical and long periods between replenishment. Such equipment needs to comply to BS5345 Part 2 and meet Health and Safety Executive requirements.

All of the above techniques and equipment are standard for water treatment works application. The modification of such designs to suit secondary chlorination is nevertheless far from straighforward and requires a good deal of experimentation and modification before we reach the stage in the UK where optimum secondary chlorination can easily be applied.

CONCLUSIONS

Reflecting the recently more stringent implementation of drinking water quality standards, a trend can be anticipated in the UK of improved disinfection control at sourceworks and the installation of many more secondary chlorination stations within distribution.

Presently, the design of disinfection systems is related to the achievement of microbiological quality objectives but this approach on its own can be compromised by the uncertainties arising from low sampling frequencies; further, various microbiological issues have tended to be over-looked or are in need of development.

A more objective basis is proposed, for managing the concentration of chlorine residuals within distribution, which would lead to a more consistent approach to achieving microbiological standards.

Optimisation of secondary chlorination will only be achieved given a more objective approach and the pursuance of investigations to ensure that correct doses are applied at correct locations.

REFERENCES

Burke V, Robinson J, Gracey M, Peterson D and Partridge K (1984)
    Isolation of Aeromonas hydrophila from a metropolitan water
    supply: seasonal correlation with clinical isolates. App.
    Environ. Microbiol. Vol 48, No 2, 361-366.

DoE (1983). "The bacterial examination of drinking water
    supplies 1982", Department of the Environment / Department
    of Health and Social Security / Public Health Laboratory
    Service, HMSO.

DoE (1984). Letter WI7/1984

DoE (1989). Group of Experts on Cryptosporidium in Water Supplies
    Interim Report of 25th July 1989.

DoE (1989b). Guidance on safeguarding the quality of public water
    supplies, Department of the Environment, HMSO.

EC (1980). Council Directive of 15 July 1980 relating to the
    quality of water intended for human consumption, Official
    Journal L229.

Galbraith N S, Barrett N, Stanwell-Smith R (1987). Water
    associated disease after Croydon, Inst. Wat. Eng. Sci.
    Summer Conference.

Schellart J A (1987). Chlorination and drinking water quality:
    Amsterdam drinking water quality before and after stopping
    safety chlorination, Inst. Wat. Eng. Sci. Summer Conference.

Schroders (1989). Prospectus for the Water Share Offer.

WAA (1988). Operational guidelines for the protection of
    drinking water supplies, Water Authorities Association,
    London.

WHO (1984). Guidelines for drinking water quality, World Health
    Organization, Geneva.

WRc (1986). Viral surveillance in the water industry: needs and
    feasibility, Water Research Centre Seminar, London.

DISCLAIMER

The views expressed in this paper are those of the Authors and
do not necessarily relate to the views of Water Quality
Management Ltd.

Chapter 14

MEASUREMENTS OF MASS TRANSFER IN A BUBBLE COLUMN WITH DIFFERENT HYDRODYNAMIC CONDITIONS

C Lang and G Bachmeier (Institute of Hydromechanics, University of Karlsruhe, Fed. Rep. Germany)

ABSTRACT

The effect of hydrodynamic conditions in the liquid phase of a vertical, co-current air water bubbly flow on the mass-transfer of oxygen has been investigated experimentally. It is discussed which fluiddynamic mechanisme have to be considered to describe the behaviour of such an air-water flow. The parameters turbulence intensity and mean flow velocity of the water flow and the air flow rate were varied and correlated with the overall volumetric mass-transfer coefficient $k_L \cdot a$.

NOMENCLATURE

A	interfacial area through which the oxygen transfer occurs $[L^2]$
a	interfacial area A per unit volume of liquid V $[L^{-1}]$
b	bar width of the turbulencegrid $[L]$
C_L	O_2-concentration in the bulk liquid $[ML^{-3}]$
	Indices 0 - at the time t=0
	t - at the time t
C_S	equilibrium O_2-concentration (corresponding to Henry's law) $[ML^{-3}]$
D	diameter of the bubble column $[L]$
D_L	molecular diffusion coefficient $[L^2 T^{-1}]$
d_B	diameter of bubble $[L]$
$k, k_L \cdot a$	overall volumetric mass-transfer coefficient $[T^{-1}]$
$\bar{k}_L$	transfer coefficient of the liquid phase $[MT^{-1}]$
L_y	horizontal macroscale of turbulence $[L]$
M	mesh length of the turbulencegrid $[L]$
$\dot{n}$	gas flux $[MT^{-1}L^{-1}]$
Q_a	air flow rate $[L^3 T^{-1}]$
Q_w	water flow rate $[L^3 T^{-1}]$
R	correlation coefficient $[-]$
Re	$(U_{SL} \cdot D)/\nu$, Reynoldsnumber $[-]$

Water Treatment – Proceedings of the 1st International Conference, pp. 141–154

r	rate of the renewal of the interface $[T^{-1}]$
T	temperature $[^0C]$
Tu	turbulence intensity $[-]$
t	time variable $[T]$
U_{SG}	$Q_a/(\pi \cdot D^2/4)$ spatial mean velocity of the gas phase $[LT^{-1}]$
U_{SL}	$Q_w/(\pi \cdot D^2/4)$ spatial mean velocity of the liquid phase $[LT^{-1}]$
u_L', v_L', w_L'	r.m.s. value of the velocity fluctuations $[LT^{-1}]$
$\bar{U}_L$	mean velocity of the liquid phase in the x direction $[LT^{-1}]$
V	total volume of the liquid $[L^3]$
x	longitudinal coordinate $[L]$
x'	distance downstream from grid $[L]$
y, z	transversal coordinates $[L]$
β	Q_a/Q_w, void fraction $[-]$
δ_L	thickness of the interfacial boundery layer of the liquid $[L]$
η	$(\xi_0 - \xi_{E0})/\xi_0$, efficiency of oxygen transfer $[-]$
ν	kinematc viscosity of the liquid $[L^2T^{-1}]$
ξ_0	volumetric O_2-concentration in the air input $[-]$
ξ_E	volumetric O_2-concentration in the air output $[-]$

Indices 0 - at the time t=0

　　　　　 t - at the time t

1. INTRODUCTION

Gas-liquid mass-transfer plays a role in a variety of problems in chemical and biological engineering [16], [13]. In particular, aeration in water and waste water treatment or desorption of volatile organic compound in ground water are some of the areas of research that are concerned with the problem of mass-transfer. In both cases, aeration-as well as stripping-processes, the purpose is the creation of appropriate interfacial areas required for the transfer of the water and/or air contents. Thus, aeration of bubbles is often used to increase gas-liquid mass transfer in such adsorption and/or desorption processes, see Fig.1.

The efficiency of these exchance processes is substantially influenced by the hydrodynamics of the two-phase flow. In such a system of gas and water mixture, the phase boundaries are deformable and therefore by nature unsteady. A detailed analytical description of this phenomena with often unknown influxes is not available. Planning and dimensioning of aeration or stripping systems require however not only knowledge of the influence of fluid flow mass-transfer, but also knowledge about the realization of the desired flow conditions by appropriate reactor design and construction.

To clarify the basic correlations between mass-transfer and fluid flow, and in order to quantify the overall volumetric mass-transfer coefficient k, the detailed processes in the gas and fluid flows are being investigated. The parameters, turbulence intensity and structure, bubble size and the duration of contact, are of main interest.

A bubble column was built with a special air input system (see Fig. 3, 4) to produce single bubbles with diameters from 0.2 to 10 mm and bubble swarms. The influence of the above mentioned parameters on the mass-transfer coefficient was determined by measurement of oxygen transfer.

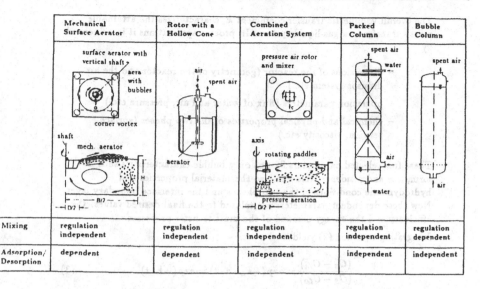

	Mechanical Surface Aerator	Rotor with a Hollow Cone	Combined Aeration System	Packed Column	Bubble Column
Mixing	regulation independent	regulation independent	regulation independent	regulation independent	regulation dependent
Adsorption/ Desorption	dependent	dependent	independent	independent	independent

Fig. 1: Adsorption/ Desorption Devices

2. METHODS/EXPERIMENTAL PROCEDURE

2.1 Basic Theory

The transfer across a gas-liquid interface of gases of low solubility (such as oxygen) is controlled by resistance of the liquid side [17]. Hence it is the interaction of processes of molecular diffusion and the liquid turbulence structure near the interface that determines the gas flux, which may be characterized by the liquid transfer coefficient k_L. The gas flux $\dot{n}$ specifying the rate at which the gas volume M (oxygen) transported across interface area A from the bubble to the bulk liquid (water), see Fig. 2, may then be related to gas concentration as

$$\dot{n} = \frac{dM}{dt} = k_L \cdot A \cdot (C_S - C_L) \tag{1}$$

Chemical reactions between gas and fluid are excluded. The change of the oxygen concentration in water by unit time gives the following equation [20]:

$$\frac{dC_L}{dt} = \frac{dM}{dt} \cdot \frac{1}{V} = k_L \cdot a \cdot (C_S - C_L) = k \cdot \Delta C \tag{2}$$

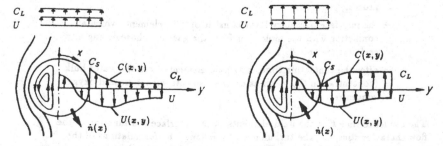

Fig. 2: Mass-Transfer across the Interface

The overall volumetric transfer coefficient k, characterizes the attainable mass - transfer in a gas-liquid system. In process applications it is influenced by

- dimensions of the reactor (geometry of the reactor and the air diffusor system etc.)
- operation parameters (flux of water and air, pressure etc.)
- chemical and physical properties of the two phases (surface tension, viscosity etc.)

These free selected parameters fix not only bubble characteristics such as form, size and residence time but also the material properties and hydrodynamic conditions of the two phases and the interface boundary layer. Now these dependent parameters are related to the final desired values: the efficiency and the energy demand of the mass-transfer processes.

Integrating equation (2) yield:

$$\frac{(C_S - C_{Lt})}{(C_S - C_{Lo})} = exp(-k_L \cdot a \cdot t) = exp(-k \cdot t) \qquad (3)$$

This equation corresponds to a kinetic reaction model of first order and leads to the determination of k.

2.2 Overall volumetric mass-transfer coefficient k

For practical transfer processes, it is not possible to identify separately the two values k_L and a. To dimension and regulate a mass transfer reactor, it is of great importance, knowing the parameters which influence these values, to take suitable engineering actions for their variation. For this, it is necessary to clarify the physical conditions in the neighbourhood of the interfacial boundary layer.

The replacement of the liquid in the region of the interfacial boundary, is often described in terms of renewal models. A comparison of the various theories and definition equations for the transfer coefficient k_L is listed in table 1.

All formulate k_L as a function of the terms:

$$k_L = f(D_L, r, \delta_L) \qquad (4)$$

The following assumptions are common to all models:

- the oxygen diffusion through the interfacial area is a molecular process,
- the interface boundary layer consists of fluid elements which are contacting with the molecules from the gaseous phase during a certain time unit t',
- after the contacting time, new fluid elements from the water bulk (concentration C_L) push aside the elements with increased oxygen concentration.

The residence time t' of the fluid elements at the interface is a function of the flow characteristics, i.e., the turbulence of the flow. The formulations of the models 6, 7 and 8 (table 1) describe this influence of the turbulent flow on the

Author	Correlation	Model	Definition	Comment
Lewis/Whitman [12] **(1)**	$k_L = D_L/\delta$	Two-Film Theory	δ_L = thickness of the liquid film	fixed interfacial area; stationary gas transfer; 1. Fick' Diffusion law
Higbie [9] **(2)**	$k_L = 2\left(\frac{D_L}{\pi \cdot t}\right)^{1/2}$	Penetration Theory	t' = residence time of the interfacial area	2. Fick' Diffusion law t' = constant
Danckwerts [4] **(3)**	$k_L = \left(\frac{D_L}{s}\right)^{1/2}$	Random Surface Renewal	s = frequency of the surface renewal	2. Fick' Diffusion law
Dobbins [7] **(4)**	$k_L = \sqrt{D_L \cdot r} \cdot \cot\sqrt{\frac{\delta_L^2 \cdot r}{D_L}}$	Continuously Exchanged Film	r = renewal rate of the interfacial area; δ_L = see model (1)	modification of models (1) and (3): $r \to 0$ see model (1) $r < 0$ see model (3)
Toor, Marchello [6] **(5)**	$k_L = f(D_L, r, \delta_L)$	Film-Penetration Theory	see model (1) and (4)	total description of models (1),(2),(3) with the distinction: t' = constant t' = incidental
Dobbins [8] **(6)**	$k_L = \frac{D_L}{\eta_K}$	Microscale model	η_K = microscale of turbulence	eddies with microscale η_K influenced the interfacial area.
Fortescue, Pearson [14] **(7)**	$k_L = C_1\sqrt{\frac{D_L \cdot u_L'}{L}}$	Macroscale model	L = macroscale of turbulence; u_L' = r.m.s. value of vel.fluct.; C_1 = constant coefficient	eddies with macroscale L and a longer residence time t' influenced the interfacial area.
Lamont, Scott [14] **(8)**	$k_L = C_2\sqrt{D_L \cdot \left(\frac{\xi}{\nu}\right)^{1/2}}$	Eddy-cell model	ξ = total dissipation rate; ν = kinematic viscosity of the liquid; C_2 = constant coefficient	superposition of big and small eddies.

Table 1: Mass-Transfer Models

transfer of the gas from the interface into the bulk of liquid by characteristic turbulent scales. The problem consists in the measurement of these scales in the two phase flow [6], [10].

For the quantitative and qualitative determination of the volumetric interfacial area a as a function of flow and material characteristics two sorts of measurement methods [6], [3], [19] are commonly used:

- chemical methods leading to a direct determination,
- physical methods analysing bubble characteristics such as local phase distribution, spatial bubble diameter distribution (e.g. Sauter diameter)

These methods necessitate a substantial disturbance of the system or are now under development.

2.3 Approach

For these investigations a bubble column was chosen because of the simple regulation of the flow rate, the flow characteristics and the air input in this system. The relative importance of turbulence in influencing oxygen transfer from bubbles to water is discussed. It is not intended to specify design parameters in this paper, but rather to emphasize the importance of this parameter. Therefore the problem is reconsidered in a much simpler situation, namely of a uniform mean flow with approximately isotropic turbulence characteristics [5]. However hypothetic such flows might seem from a practical point of view, their study is believed to be an essential first step towards the more elaborate investigation of other more realistic two-phase flows.

2.4 Experimental Facility

The measurements were conducted in a closed-circuit system with a volume V of 2 m^3 of water (see Fig. 3). The flow is driven by two pumps with variable discharges Q_w up to 0.070 m^3/s, which is equivalent to a Reynolds number of 2.0 $\cdot 10^5$. The working cross-section, made of plexiglass, measured 2 m high with a diameter of 0.44 m. A uniform velocity distribution was obtained by an arrangement of screens in a diffuser. Freestream turbulence was found to be lower than 5% (= type T0). A variation in turbulence intensity could be achieved by grids at the entrance of the cross-section. The size of the selected grid mesh is M=5 cm and the quadratical bars are b=1,5 cm per side (= grid type T1).

Gas (Air) is injected through 25 nozzles, positioned at each cross of the turbulence grid (see Fig. 4). The gas flow rate in each nozzle can be controlled individually. The mass-transfer measurements have been carried out in co-current flow with an average volumetric void fraction, β, in the range of $0 < \beta < 10\%$.

The continous measurements of dissolved oxygen in the fluid were performed by means of membrane electrodes in accordance to the ASCE-Standards [1]. Determination due to this dynamical method is permissible because of the relative short length of the column and the small fineness ratio. The assumption of complete mixing is then justified. The air was analyzed before and after passing the bubble column in an oxygen detector. Because of this method it is possible to carry out a mass-balance of the system.

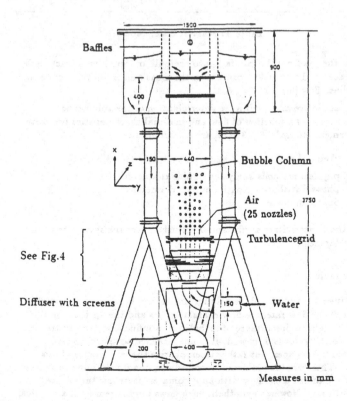

Fig. 3: Cross-Section of the Bubble Column

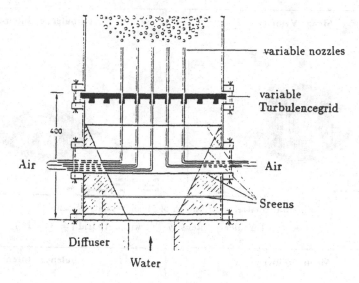

Fig. 4: Air-Input System

The velocity and turbulence measurement in the single phase flow is obtained using Laser-Doppler Velocimetry (LDV). The LDV-system was a one-component system, based on a 15 mW laser, operated in forward scattering mode with bragg-cell shifting. The probe volume is about $0.8\ mm \times 1\ mm$. The photomultiplier signal was filtered and processed by a computer. To eliminate effects of the circular wall, a transparent rectangular box containing water was mounted around the bubble column.

The experimental procedure was the following: first of all, the catalyst salt $CoSO_4$ was dissolved in the circulating water. Then slowly the oxygen binding salt Na_2SO_3 was added until the oxygen which is dissolved in the water by nature is no longer measureable. The reaeration is then started by opening the air-valve. The air supply is measured by a rotameter. All data were taken with an A/D converter at a rate of 0.1 Hz on a personal computer. The total measuring time per configuration varied from 70 - 90 min.

3. RESULTS AND DISCUSSION

3.1 Characterization of the water flow

The longitudinal and transversal coordinates are denoted as x, y, z (see Fig. 3). The main characteristics of the flow field in the absence of bubbles, namely the mean velocity $\bar{U}_L$ and the turbulence intensity Tu_x were measured without and with the turbulence producing grid at various points in the bubble column. The turbulence intensity Tu_x is defined here as the ratio of the r.m.s. value, u'_L, of the longitudinal velocity fluctuations and the mean velocity.

Typical results are shown in a 3-D representation for both grid conditions for a flow rate, $Q_w = 0.07\ m^3/s$. The measurement point is located in a distance of $3 \times$ M from the grid T1 directly above the air inlet nozzles ($x'/D = 0.09$), see Fig. 5. When the column is operated without the grid (type T0), the mean velocity is uniform within 5 %, $\bar{U}_L = 0.52\ m/s$. In addition, the turbulence intensity is similarly uniform in the same region $Tu_x = 5\% \pm 1\%$. On the other hand, with the grid (type T1), the mean velocity and the

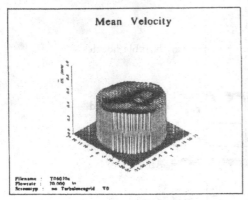

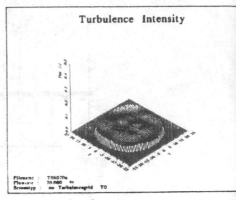

Fig. 5a: Spatial Distribution of Mean Velocity $\bar{U}_L$ and

Turbulence Intensity Tu_x without Grid (= type T0)

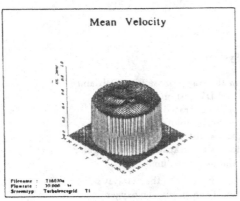

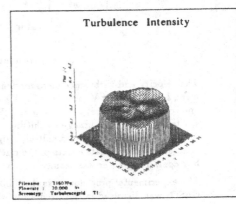

Fig. 5b: Spatial Distribution of Mean Velocity $\bar{U}_L$ and

Turbulence Intensity Tu_x with Grid (= type T1)

turbulence intensity again prove to be constant over the cross-section, but turbulence intensities now increase to a value of $Tu_x = 23\ \% \pm 2\ \%$.

Moreover, the measurements of the longitudinal and transversal velocity fluctuations $(u'_L,\ v'_L,\ w'_L\ )$ indicate the degree of isotropy of the turbulence without and with the grid (Fig. 6). This fact is described by *Baines and Peterson* (1958) for air flow and also by *Lance and Bataille* (1984) in water flow.

Finally the decay of turbulence is governed by the following asymptotic law:

$$Tu_x = \frac{u'_L}{\bar{U}_L} = 1.12 \left(\frac{x}{b}\right)^{-\frac{5}{7}} \tag{5}$$

In Fig. 7 is plotted the variation of turbulence intensities along the axis for all test conditions. No noticeable influence of the Reynolds-number, Re, in the range studied was found.

Baines and Peterson (1958) also gave a function for the lateral scale of turbulence:

$$\frac{L_y}{b} = 0.1 \left(\frac{x}{b}\right)^n \qquad 0.53 < n < 0.56 \tag{6}$$

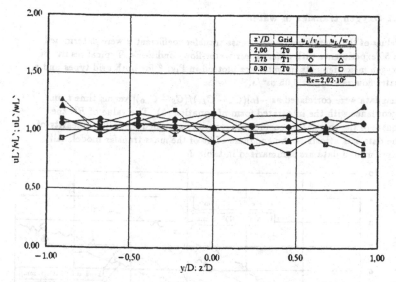

Fig. 6: Distributions of the Ratio of Fluctuations u'_L/v'_L, u'_L/w'_L in y-
and z-Direction

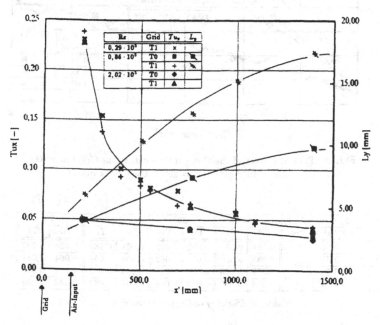

Fig. 7: Decay of Turbulence Intensity Tu_x and Variation in Macroscale L_y
of Turbulence

The lateral scale L_y is one of these values which often is used to describe the
condition of a turbulent flow. The results for the grid types are also plotted
in Fig. 7.

In summary, the turbulence conditions of the water flow in the bubble
column, characterised by the turbulence intensities and the macroscale of
turbulence can be influenced distinctly by the choice of the grid type.

3.2 Oxygen transfer in water

Values of the overall volumetric mass-transfer coefficient k were determined in 50 experiments with different air-water-flow conditions. Typical results of oxygen measurements in water are plotted in Fig. 8 for both grid types and a water flow rate $Q_w = 0,05 \ m^3/s$.

The data were correlated as $-ln[(C_S - C_{Lt})/(C_S - C_{L0})]$ versus time t, in accordance with the integrated form of equation (3). The value of R^2, the square of the correlation coefficient, generally exceeded 0.99. The linearity of the data sets attests to the goodness of fit of the mass-transfer model. All experimental data are summarized in table 3.

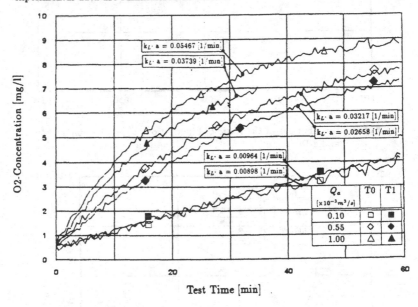

Fig. 8: Typical Curves of the Oxygen Concentrations $C_L(t)$ in water
$(Q_w = 50 \times 10^{-3} \ m^3/s, \ T = 20^{\circ}C, \ C_S = 9,02 \ mg/l)$

$Q_w \ [\times 10^{-3}m^3/s]$	10		30		50		70	
$Q_a \ [\times 10^{-3}m^3/s]$	T0	T1	T0	T1	T0	T1	T0	T1
0,10	0,086	0,083	0,082	0,082	0,084	0,082	0,084	0,082
0,30	0,083	0,080	0,066	0,062	0,068	0,067	0,064	0,061
0,55	0,079	0,068	0,051	0,051	0,054	0,052	0,051	0,050
1,00	0,074	0,060	0,050	0,046	0,047	0,042	0,044	0,039

Table 2: Efficiency of Oxygen Transfer η

$Q_w \ [\times 10^{-3}m^3/s]$	10		30		50		70	
$Q_a \ [\times 10^{-3}m^3/s]$	T0	T1	T0	T1	T0	T1	T0	T1
0,10	0,00885	0,00806	0,00887	0,00844	0,00893	0,01006	0,01124	0,01053
0,20	0,01510	0,01415	0,01368	0,01256	—	0,01359	—	0,01725
0,30	0,02084	0,01997	0,01829	0,01729	0,01961	0,01853	0,02009	0,01956
0,40	0,02501	0,02477	0,02122	0,02022	—	0,02041	—	—
0,55	0,03844	0,03251	0,03104	—	0,03217	0,02601	0,03457	0,03090
1,00	0,06611	0,04866	0,05065	0,03261	0,05300	0,03733	0,05402	0,04680

Table 3: Results of the Experiments: Overall Volumetric Mass-Transfer Coefficient $k_L \cdot a \ [1/min]$

The results for all measurement conditions and both grid types T0 and T1 are shown for the oxygen concentration in water in Fig. 9 and 10 as a function of the void fraction β and also for oxygen efficiency in air in table 2. Figure 9 shows that the influence of the grid types T0 and T1 occurs in all water flow rates Q_w. At the lowest air flow rate $Q_a = 0.1 \cdot 10^{-3} m^3/s$ the results of the oxygen measurements are approximately similar for both grid conditions ($k \sim 0.01\ 1/min$). With an increasing Q_a the difference between the k-values without and with grid also increases. For $Q_a > 0.55 \cdot 10^{-3} m^3/s$ the curves taken from the experimental data with the grid show a nearly asymptotic behaviour.

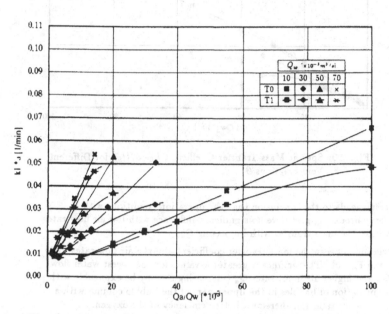

Fig. 9a: Overall Volumetric Mass-Transfer Coefficient $k_L \cdot a$ for Different Water Flow Rates without and with the Grid

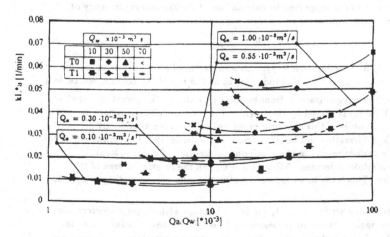

Fig. 9b: Overall Volumetric Mass-Transfer Coefficient $k_L \cdot a$ as a Function of Air Flow Rate without and with the Grid

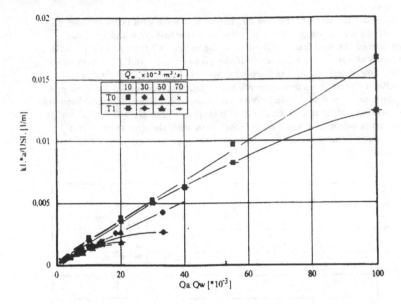

Fig. 10: Specific Mass-Transfer Coefficient $k_L \cdot a/U_{SL}$ for Different Void Fractions β

This result leads to the conclusion that the operation conditions with a turbulence producing grid have no significant influence with low air flow rates but reduces the mass-transfer with increasing Q_a.

The relation between the mass-transfer coefficient and the air flow rates are plotted in Fig. 9b. The significant greater k-values for the lowest water flow rate Q_w and higher air input rates Q_a (increasing β) may be explicable with the accumulation of bubbles in the upper part of the bubble column with a longer residence time and therefor a better efficiency of the oxygen.
The calculation of k related to the spatial mean velocity of the gas phase U_{SG} results in decreasing values with increasing air flow rates (comparison of the maximum and minimum values: Q_a– factor 10; k– factor 5).

These results are supported by calculation of the maximum efficiency of oxygen transfer:

$$\eta = \frac{\xi_0 - \xi_{E0}}{\xi_0} \tag{7}$$

where ξ_0 is the volumetric oxygen concentration of air input in the bubble column and ξ_{E0} is that of the air output at the top of the column, see table 2.

The better oxygen transfer from bubbles to water at the lowest air input Q_a ($\eta \sim 6 - 8\ \%$) is caused by the single rising bubbles with diameters $d_B = 3 - 5\ mm$. Bubbles of this diameter range enhance oxygen transfer by bubble deformation and bubble surface oszillations [15]. The movement of the bubble surface results in the continous production of of great local concentration differences $\Delta C = (C_S - C_L)$ in the interfacial area of the bubbles. This leads to an increasing mass-transfer corresponding to equation (2).

An increasing air flow rate Q_a produces bubbles with larger diameters and bubble agglomeration by coalescence. This effect leads to reduction of the interfacial area and to a adequate loss in oxygen entry [15].

A plotting of the specific mass transfer coefficient $k_L \cdot a/U_{SL}$ considers the influence of spatial mean velocity of the liquid phase U_{SL} and also the residence time of the bubbles. Figure 10 shows that the largest values can be reached with a low water flow rate $Q_w = 0.01 \ m^3/s$. The influence of the turbulent single water flow conditions is also seen in this figure.

4. CONCLUSIONS

Overall volumetric mass-transfer coefficients $k = k_L \cdot a$ and maximum efficiency of oxygen transfer η for air bubbles rising in water with different hydrodynamic conditions have been measured in order to quantify oxygen mass-transfer. The results of the experimental measurements suggest the following conclusions: Mass-transfer depends on

- the air flow rate Q_a influencing bubble diameter as well as bubble deformation and residence time,

- the water flow rate Q_w mainly affecting the residence time of the bubbles

- the turbulent conditions of the water flow .

To clarify these effects and to assess the importance of each of them further work is necessary. Currently, experiments are prepaired with a variety of turbulence grids, producing different turbulence conditions in the water flow. At present the influence of the differnet hydrodynamic conditions on the bubble diameter distribution, produced by the air-input system with variable nozzels is investigated. For this photographic and opto-electronical methods are used.

ACKNOWLEDGEMENTS

This work was supported by the Bundesministerium für Forschung und Technologie (BMFT), Project number: 02-WA 8680/4

BIBLIOGRAPHY

[1] **ASCE** (1984): A Standard for the Measurement of Oxygen Transfer in Clean Water. ASCE, Vol.6.

[2] **Baines, W.D.; Peterson, E.G.** (1951): An Investigation of Flow Through Screens. ASME, Vol.7, 467-480.

[3] **Brauer, H.** (1971): Mass Transfer. Sauerländer, Frankfurt.

[4] **Danckwerts, P.V.** (1970): Gas-Liquid-Reactions. McGraw Hill, New York.

[5] **Davies, J.T.** (1972): Turbulence Phenomena, Academic Press, New York.

[6] **Deckwer, W.-D.** (1985): Technical Reactions in Bubble Columns. Salle + Sauerländer, Frankfurt.

[7] **Dobbins, W.E.** (1956): The Nature of the Oxygen Transfer Coefficient in Aeration Systems. Treatment of Sewage and Industrial Wastes, McCabe.

[8] **Dobbins, W.E.** (1962): Mechanism of Gas Absorption by Turbulent Liquids. Water Pollution Research, Pergamon Press, Vol. II.

[9] **Higbie, R.** (1935): The Rate of Absorption of a Pure Gas into a Still Liquid during Periods of Exposure. Trans. Am. Inst. Chem. Eng., Vol.31, 365-389.

[10] **Lance, M.; Bataille, J.** (1983): Turbulence in the Liquid Phase of a Bubbly Air-Water Flow. NATO ASI Serie E-No. 63, Martinus Nijhoff Publishers.

[11] **Lance, M.; Marie, J.L.; Bataille, J.** (1985): Homogeneous Turbulence in Bubbly Flows. Fundamental Aspects of Gas Liquid Flows. ASME, New York.

[12] **Lewis, W.K.; Whitman, W.G.** (1924): Principles of Gas Absorption. Industrial and Engineering Chemistry, Vol.16.

[13] **Matter-Müller, Ch.; Gujer, W.; Giger, W.** (1981): Transfer of Volatile Substances from Water to the Atmosphere. Water Research, Vol.15, 1271-1279.

[14] **Mersmann, A.** (1986): Mass-Transfer. Springer, Berlin.

[15] **Nestmann, F.** (1984): Oxygen Entry by Bubbles and Development of a Mechanical Aeration Device. Thesis, University of Karlsruhe.

[16] **Nestmann, F.; Lang, C.** (1986): A Mechanical Device for Separate Mixing and Aeration in Sewage Tanks. Wasserwirtschaft, Vol.2, 52-57.

[17] **Treybal, R.** (1981): Mass-Transfer Operations, McGraw Hill, New York.

[18] **Sengpiel, W.; Meyder, R.** (1987): An Experimental Investigation of Local Void Distribution and Turbulence Structure of Bubbly Two-Phase Flow in Vertical Channels. 4th Workshop on Two-Phase Flow Predictions, Erlangen.

[19] **Wallis, G.B.** (1969): One Dimensional Two-Phase Flow. McGraw Hill, New York.

[20] **WPCF; ASCE** (1988): Aeration. ASCE-WPCF Manual and Report on Engineering Practice, No.63.

PART V

Treatment: innovation and application

Chapter 15

WATER TREATMENT TECHNOLOGIES FOR THE CHALLENGES OF THE NINETIES

F Fiessinger (Zenon Environmental Inc., Burlington, Ontario, Canada)

INTRODUCTION

In early 1985, I wrote with a group of researchers from Lyonnaise des Eaux Research Center, a general paper (1) on where we felt the technologies of water treatment were heading on to. This paper was first presented at a workshop on "Environmental Technology Assessment" held at the University of Cambridge in England on April 24, 1985. It was reproduced at the end of 1985 in the "Water Research Quarterly" of the AWWA Research Foundation and through this it got an unexpected but considerable publicity throughout the United States. The title of the paper but also its content had been greatly influenced by the famous book and # 1 bestseller, Megatrends of John Naisbitt (2). This book was identifying trends and pressures which would likely force the world to change.

Five years later the trends have not significantly changed and this "global" approach still appears the most sensible for a keynote paper. Rather than attempting to predict the emergence of specific technologies, I would like, in this general presentation, point out probable changes which may occur and draw from these the potential emergence of new technologies. It does not pretend to be the review of all foreseeable evolutions, but rather a presentation of a few trends in water treatment technologies, which seem most likely to develop in the course of the coming decade.

THE CHALLENGES OF THE NINETIES

All around the world, the water sector is undergoing accelerating changes. These changes are due in part to the strategy of the organizations themselves. They are however, primarily created by external forces beyond their control. Ten of these forces can be easily identified (4). Their importance varies considerably from one country and even one region to another, the list is not comprehensive and each of them may interact with the others.

i)   The reinforcement of quality standards accompanied by an increasing demand for high quality water and a cleaner environment from the general public.

ii) The emergence of stronger ("greener") pressure groups.

Water Treatment – Proceedings of the 1st International Conference, pp. 157–164

iii) The progress in technology, such as electronics, biotechnology, polymer science, membrane separation, etc...

iv) The spin-offs from the development of treatment of water for specific applications, such as ultra-pure water for laboratories, water for the electronic industry, for enhanced oil recoveries, bottled water etc...

v) The quantitive and qualitive reduction in water resources. This is sometime related to pressure for a better environment.

vi) The need to recycle water (e.g. in the buildings in Tokyo).

vii) The economic pressure exerted to reduce energy, labour costs and increase productivity.

viii) The development of privatization resulting in increasing competition between water organizations. This forces public services to look for new markets and therefore develop a policy of diversification. The most common one is water treatment. This in turn will result in the development of "large" water organizations.

ix) The emergence of more and more innovative and broader flexible financing and operating options for water facilities.

x) The changes in the mentality of waterworks personnel: No more "hierarchy", but more team work, etc...

Today, more than ever it seems that in this list the strongest trends enhancing changes in water treatment are the following:

• The need for quality and the new U.S. regulations.

• The progress in technology and in turn the demand for "high-Tech",

• The development of privatization more particularly in Europe, where the completion of the single market after 1992 might accelerate changes.

The conjunction of several of these forces might literally create "window of opportunities" for specific technologies in the coming few years and it is likely in fact that most of the changes will occur in the early part of the 1990's.

THE RESEARCH ORIENTATIONS

The second part of the previous decade has seen a rise (4) in R&D expenditures, and the increasing involvement of the water utilities themselves in carrying out this research. In these expenditures, R&D in water treatment, usually represents more than one third and frequently one half of the total. In the coming decade it is thus expected that water treatment will reap the fruits of these efforts.

Another observation in the trends to change is the fact that after several years of work more focused on wastewater treatment, it appears that the drinking water investigations are now becoming more important. This is probably related to the influence of the new regulations such as the U.S. "Safe Drinking Water Act" (SDWA). Industrial wastewater treatment remains also of great importance and its high specificity (see also vi above) triggers sophisticated technology spin-offs.

The bulk of the R&D in the drinking water area, however, seems to be rather conservative: it focuses more on a better understanding of the intimate mechanism and a better control of existing processes than on the development of new ones. A large portion of the effort is also spent in pilot experiments, which are usually local demonstrations, of a proven technology. As an example among the papers selected in the journal AWWA for their 1989 best paper award,30% were on activated carbon absorption, 25% were on ozone oxidation,10% were on filtration, 10% were on membrane separation, 10% were on air stripping, and 10% were on sludge dewatering.

These papers represent research projects which were probably started several years ago and the current reality may be somewhat different. We can see, however that in the area of drinking water, carbon absorption and ozone oxidation still represent more than half of the research efforts. This is a rather general situation, but new technologies such as membrane separation are now appearing heavily in the programs and some research organizations are deliberately putting the majority of their strengths on these innovative areas.

THE DRINKING WATER AREA: A FOCUS ON ORGANIC MATTER REMOVAL

It is needless to say, that water treatment will, in the years to come, continue to focus on an improvement of the removal of the organic constituents. The push for reinforced standards following the U.S. SDWA, the need to eliminate all the new synthetic organics appearing in the environment, the attempt to avoid the formation of disinfection by-products, the requirement to control the bacterial regrowth in the distribution systems,... All these reasons will lead to technologies more and more effective in organics removal. This does not mean of course that local problems caused by other substances may not arise. Compounds such as Radon, heavy metals, radioactive metals, barium and of course calcium, sodium, potassium. will always generate treatment technology improvements. The main thrust will however remain on the organics, for the years to come.

Together with treatment technologies tremendous efforts and progress will be made in the areas of water analysis, identification of the mechanisms of formation of the by-products, health effects assessment and the next decade might see the development of toxicity bio-assays.

ABSORPTION, OXIDATION AND DISINFECTION

These treatment technologies which are traditionally being used for the reduction of organic matter will naturally continue to grow in importance. A better usage of the granular activated carbon (GAC) through a better modelling of the absorption mechanisms seems to be the surest evolution of current practice. Alternatives to GAC such as activated alumina, which raised some interest in the mid eighties is now fading out. New carbons, with various shapes (fibers?) might appear, better reactivation technologies will develop, but altogether this will result only in minor changes over the technologies which have been in use for the past fifteen years. Powdered activated carbon (PAC) will probably regain importance over GAC, particularly in conjunction with floc blanket clarifiers and new membrane reactors.

The development of ozone will continue. It will be more and more systematically applied together with GAC, but again, this is not a really new trend!

The technology of its generation and of its contacting with water will consistently be improved. The newest trend is probably in the development of the combination of ozone with other oxidants such as U.V. and H_2O_2. New reactors allowing an efficient combination of these oxidants will in parallel, be developed.

Chlorine will remain as a safety disinfectant, at very low dosages, after removal of the core of the organics and of all potential precussors of THMS. Chlorine dioxide will probably remain of limited usage and will not undergo any significant development over the existing situation.

BIOLOGICAL TREATMENT

It is still hoped that future water treatment technologies, even in the drinking water area, will make a greater use of biological processes. Nitrification and more particularly denitrification processes will be developed. The removal of trace organics by the nitrifying biofilm, the development of biofilters for metal precipitation, will all undergo interesting developments. The control of the biofilm in the distribution system, where disinfectants will be at trace levels will be of major concern and will force the use of some form of "biofilters" at the plant itself to get biostability. No significant improvement, however, seems likely to occur and although the steady rise in research efforts in this area will continue, the traditional weaknesses of biological treatments i.e. lack of reliability and strong sensitivity to flow and loadings changes will cripple the expansion of their application to drinking water production. The development of the membrane-bioreactors, however may bring about unexpected breakthroughs in this area as well.

CLARIFICATION

A few years ago, I expected an important change in this area due in particular to the arising of the polymeric coagulants. It did not occur and it is more and more unlikely that it will! The quest for quality renders more and more difficult the use of any chemical. The replacement of conventional inorganic coagulants such as alum and ferric chloride by more efficient polymeric species, is progressing very slowly. In parallel rising suspicions on the relation between aluminum residual and the Alzeimer's disease may condemn the development of the use of this metallic coagulant.

Clarifiers and sand filters technology seems to have reached a plateau and no new development is likely to occur. A lot more research work will be done however in this area because of the new U.S. regulations on the filtration requirement and the guideline turbidity (0.5 JTU). This regulation will undoubtedly open a window of opportunity for membrane filtration.

MEMBRANE SEPARATION

Among all the water treatment technologies, membranes seem to be the most promising ones. Their best advantage is their ability to produce water with a constant and very well adjusted quality. At a time where the demand for quality is a major drive, membrane have thus a clear advantage on all other technologies. In addition membranes offer a variety of other advantages. They remove a wide range of substances, from particles to ions including bacteria and viruses.

Theoretically they could remove everything. They can operate without chemical addition to water, are reliable, compact and easy to automate. Their major disadvantages are still their rather high capital and operating costs, the fact that they are prone to fouling which requires often a high level of pretreatment and regular chemical cleaning. They may also have a rather important reject stream whose disposal may create problems.

Membranes can be made in a wide variety of shapes: fibers, tubes, flat sheets in the form of spiral wound modules or plate and frames. They can have an almost unlimited range of porosity - or molecular weight cut-offs - and the development of synthetic organic chemistry gives hopes of great improvement in the membrane composition. This will certainly result in lower operating pressures, decreased fouling, better resistance to disinfectants, biodegradation etc... and altogether a much longer lasting period. The new composite membranes with sophisticated coatings have now little to do with the original cellulose acetate membranes and the progress will likely accelerate in the coming few years.

Several outside forces are pulling the membrane technology toward drinking water applications. The first and primary force is the development of the market. Desalination plants using reverse osmosis membranes are being built around the world. In addition an important market is being developed in Florida where plants for a total capacity of roughly 200 MGD are already in operation and where at least an additional 100 MGD will be built every year. Membranes in Florida are not only used for desalination but more and more for organics removal (THMFP's) together with softening. Looser membranes as opposed to conventional R.O. - in the range of nanofiltration and ultrafiltration - are thus being installed and make the treatment more and more cost effective as compared to conventional coagulation flocculation and lime precipitation. Total costs for a plant of 0,5 MGD with membranes are in the range of 1.5 $/1000 gallons as compared to more than 2. $/1000 gallons for lime softening. This difference decreases, however, when the size of the plant increases. Membranes are now moving northward and westward throughout the United States, but Florida will remain indeed the nest for membrane proliferation in the drinking water sector. Altogether it is predicted that the total market for membranes in water treatment will be in the range of $500 million in year 1996.

Another reason for the progresses of membrane technology is the development of considerable international research programs such as the Aquarenaissance program in Japan and the Eureka program in Europe and several projects funded by the AWWA research foundation in the U.S. The "Eureka" program in particular, fuels the development of ultrafiltration and microfiltration membranes at Lyonnaise des Eaux, which are capable of meeting the costs of conventional filtration technologies with a whole load of technical advantages. Plants are already in operation and if these membranes can take advantage of the window of opportunity opened by the Surface Water Treatment Rule of the SDWA in the USA. They would bring then one of the most important revolution in water treatment practices.

An interesting feature of membranes is in their ability to constitute a reaction by combining the separation with a reaction in the recirculation loop. A whole world of membrane reactors can thus be envisioned: absorption reactors with powdered activated carbon in MF membranes, oxidation reactor, carbonate precipitation and iron precipitation reactors. The best hope being in the membrane/bioreactors which would significantly open the door to industrial applications of biotechnology to water treatment.

A last point of potential development is the system where membranes are included. Membranes are too often studied independently. They are essential, but only small component of an overall system which leaves considerable room for improvement. The "system" approach will likely result in great progress.

AUTOMATION

Artificial intelligence but also other forms of computer architecture such as in particular neural networks, will bring tremendous changes to all forms of water treatment technologies. They will constitute powerful tools capable of handling at a fantastic speed fuzzy data and more than anything capable of automatic learning. They may bring back to life obsolete technologies and will drastically boost the development of the most sophisticated ones. It seems that all water plants will be equipped in the course of the next decade with some form of these knowledge based systems. In parallel the development of new sensors using miniature electronics technology will be accelerated.

THE WASTEWATER AREA

The wastewater area includes primarily the development of new biological processes. It is indeed an arbitrary simplification, since a lot of the industrial wastewater treatments are primarily physical-chemical in nature and their evolution will be similar to that identified above. The main trends in the biological treatments, however, can be summarized as follows:

- A gradual change from activated sludge to biofilters.
- A pressure to develop anaerobic treatment in areas where energy cost is high (Europe, Japan,...).
- The reinforcement of phosphorus and nitrogen removal.
- The development of combined physical-chemical and biological treatments.

In this latter case we can see the potential and the probable emergence of the combination of biological treatments with membrane separation where they will be used to develop fully integrated **membrane bioreactors**. Progressively membranes will be applied in the place of secondary clarifiers. This is already taking place in the "building-plants" in Japan (recycling of municipal waste water, and in specific industrial waster water applications, requiring a high degree of reliability). Ultimately membranes (loose microfiltration) will move also to replace the primary clarifiers and will end up in making possible the development of bioreactors confined and highly specific. This may seem a little futuristic but again under the pressure of the technology development and the large international research programs, these reactors may be commonly used well before the turn of the century. They will bring about: compactness higher yields, reliability, easiness in operation, low sludge production and altogether a much better quality water.

CONCLUSION

After many years of conservation, the change in water technology has now started and will probably develop exponentially. The world itself is changing and new challenges will emerge in the coming decade which will in turn require even newer technologies. In this universe of moving targets there are some certainties: the quest for a better quality water is undoubtedly one and membranes, because of their reliability in this aspect and also because of their evolutionary nature, represent our best hope to deal with this future.

REFERENCES

1) F. Fiessinger, J. Mallevialle, A. Leprince and M. Wiesner
2) Warner books - 1984
3) John Naisbitt and the Naisbitt Group, The Years Ahead 1986, Warner books 1985
4) François Fiessinger, Applied Research in Water Supply, IWSA Rio de Janeiro, September, 1988

Chapter 16

COLLECTION EFFICIENCY OF LIQUID/LIQUID HYDROCYCLONES

J Woillez (Centre d'Etudes et de Recherches de Grenoble, France)

SUMMARY

When reduction of weight and size of oily water treatment installations is needed, liquid/liquid hydrocyclones can be used instead of settlers. An analytical model is proposed for the prediction of the collection efficiency of hydrocyclones adapted to the treatment of light dispersions. This model is derived from flow velocity measurements in a cylindrical hydrocyclone using LDA and from modelling of the effects of turbulence. Agreement with experimental data obtained with oil in water dispersions is good.

NOMENCLATURE

a : Acceleration (m/s2)
b : Correction factor
C : Light dispersion concentration(Kg/m3)
d : Oil droplet diameter (m)
$\bar{d}$: Mean inlet droplet diameter (m)
D : Cyclone main diameter (m)
E : Cyclone efficiency
f : Friction factor
K : Swirl decay parameter (m-1)
L : Cyclone length (m)
n : Exponent of vortex law
n_i : Number of droplet of diameter di
Q : Main flow rate (m3/s)
q : Extraction flow rate (m3/s)
r : Radial distance from cyclone axis (m)
R_0 : Cyclone main radius (m)
Re : Droplet Reynolds number
t_0 : Unitary shear stress at the wall (N/m2)
U : Axial flow velocity (m/s)
V : Tangential velocity (m/s)
V_0 : Tangential velocity at the wall (m/s)
V_e : Inlet velocity (m/s)
V_s : Tangential velocity at the end of the separation chamber (m/s)
W : Droplet migration velocity (m/s)
x : Axial distance along cyclone axis (m)
μ : Water viscosity (Kg/m.s))
δ : Water density (Kg/m3)
$\Delta\delta$: Density difference between light and heavy liquid (kg/m3)

Water Treatment – Proceedings of the 1st International Conference, pp. 165–172

1 INTRODUCTION

On waste water treatement installations, found for example in refineries, offshore platforms, deep well injections or spilled crude oil recovery processes, de-oiling operations are needed and are usually made by conventional settlers. But if reduction in size and weight of installations is required, this operation can be performed by hydrocyclones. It is known indeed that hydrocyclones (Ref.1) or vortex separators (Ref.2) can ensure primary and sometimes secondary separation within very short residence times of the mixtures in the system, typically a few seconds instead of a few minutes for settlers. Moreover, cut size diameters (i.e diameter of drops beyond which all drops are separated) can be smaller with cyclones than with settlers. Nevertheless, due to the complex hydraulic phenomena, the design or scaling up of these cyclones remains essentially empirical. The purpose of this paper is to propose and validate an analytical method for the design of light dispersion hydrocyclones.

2 EXPERIMENTAL SETUP

2-1 Cyclone geometry

The cyclone designed and tested by the C.E.R.G is shown on Fig.1. The flow enters the cyclone through two tangential inlets fed by an inlet chamber. The inlet velocity Ve is set to avoid drop breakup by shear stress. A converging section accelerates the flow so that the tangential wall velocity V0 at the beginning of the cylindrical separation chamber reaches a higher value V0=1.8Ve. This converging section is equipped with an internal cone which can be shown to stabilize the vortex created in the separation chamber and to generate effective high values of V0. The cylindrical separation chamber is ended by two concentric outlet orifices. The main orifice, diameter D/3, collects the clean water. The secondary axial orifice, diameter D/10, collects the oil gathered on the axis of the separation chamber. The length L of the separation chamber is 0.6m.

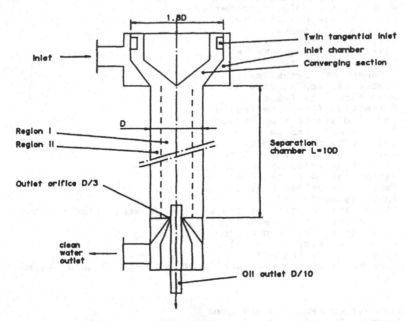

FIG 1 : Cyclone geometry

2-2 LDA measurement

A 4 W argon laser was used with the back scattering technique to measure the tangential velocities inside the separation chamber. For this purpose, the cylindrical separation chamber was shaped in a plexiglas block with parallel faces having the same refractive index as water. The optical correction to be brought to the position of the measured point and to the value of the interfringe distance was thus of low importance. The water was seeded with aluminium powder and it was assumed that the measured mean tangential velocity of these particles would be similar to that of water.

2-3 Oil in water test loop

The loop used for the oil in water separation tests is shown on Fig.2.

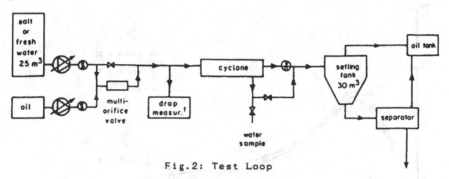

Fig.2: Test Loop

A dispersion is generated upstream of the cyclone by mixing oil in water coming from separate tanks. The mixing is operated through a specially designed valve allowing the control of the mean droplet diameter of the dispersion. The clear water outlet and oil outlet flows of the cyclone are sent to a settling tank which is filled while the tests are run and emptied before each serie of experiments.

The drop size spectra are measured by a Malvern 2600D analyser, the opticall cell of which is refilled by a continuous sampling of the main flow. A typical drop size spectra is given on Fig.3. The inlet and clean outlet oil concentration, respectively C_i and C_o, are measured from samples analysed by the tetrachloride extraction and infrared absorption method.

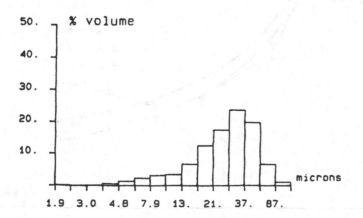

Fig.3: Drop Size Spectra d = 20 microns

3 LDA RESULTS AND ANALYSIS

Tangential velocity profiles have been measured inside the separation chamber at three locations along the x axis and for two inlet velocities. The results are shown on Fig. 4 and 5. The main conclusions which can be drawn from this results are following:

- Swirl decay is an important phenomena which affects the flow field both near the wall and in the core of the vortex. Near the wall, it can be seen that the velocity decreases from an effectively high value V0 at the begenning of the chamber to a lower value Vs at the end of it. In the core region, the maximum tangential velocity is found to be constant all along the chamber, that is, for each cross

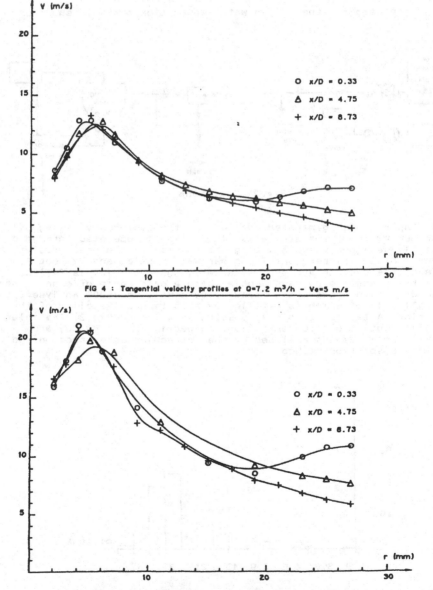

FIG 4 : Tangential velocity profiles at Q=7.2 m³/h - Ve=5 m/s

FIG 5 : Tangential velocity profiles at Q=10 m³/h - Ve=7 m/s section,it is independant of the velocity at the wall. Thus, this maximum value follows a classical vortex law only at the end of the chamber where it can be written:

(1) $V(r) = V_s.(R_0/r)^n$ with n = 0.7

- The separation chamber can be separated in two regions:
 Region I is defined by $0 < r < R_0/2$. In this region, the velocity follows a vortex law and is thus characterized by a high acceleration field.
 The other region, Region II, is defined by $R_0/2 < r < R_0$. In this region the velocity varies slowly between V_0 at the wall and $V(r=R_0/2)$, that is, according to (1), $V=1.6V_s$. Region II represents 75% of the cross section of the separation chamber and can be considered as the dominant region for the separation process in the cyclone.

4 COLLECTION EFFICIENCY

4-1 Basic principle

The basic principle of the modelling which is proposed here has been given by DIETZ (Ref.3) as used for the evaluation of collection efficiency of dust cyclones. Where due to turbulence, the radial concentration C of the dispersion is uniform at any cross section while along the separation chamber the axial concentration decreases due to the migration of droplets toward the axis.
 Here, the separation chamber is divided in two regions, I and II, defined above. In an arbitrary cross section, droplets are uniformly spread in region II and, in a given time, some of them reach region I. In this region, the acceleration field is so high that these droplets are considered to be separated from the main flow and collected along the axis of the cyclone. A time increment later, the cross section is located farther in the separation chamber and the remaining droplets are uniformly redistributed by the effects of turbulence. The above mechanism is repeated until the cross section has reached the end of the separation chamber.

4-2 Swirl decay

The modelling of the separation process described above needs a definition of the swirl decay. For this, we can write the conservation of angular momentum between two arbitrary cross sections of the separation chamber separated by a distance dx, which gives:

(2) $\delta.Q.R_0.(V_0(x+dx)-V_0(x)) = - t_0.2\pi.R_0^2 . dx$

where t_0 is the unitary stress at the wall. We can write t_0 as

(3) $t_0 = f.1/2.\delta.V_0^2$

where f is the friction factor. Putting (3) into (2) gives the differential equation

$dV_0/V_0^2 = -(\pi.f.R_0/Q).dx$

Integration for x = 0 to a given section gives

(4) $V_0(x) = V_0 /(1+Kx)$ with $K = f.\pi.V_0.R_0 / Q$

Fig.6 shows that for a classical value of the friction factor for smooth walls f =.0055, the above expression Eq.(4) gives satisfactory agreement with measured values.

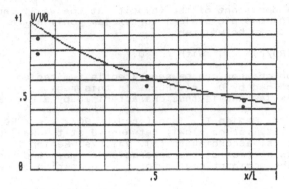

Fig.6 Swirl decay
------- Formula (4); K=2.2
Exp. points: □ Q=10.8m3/h;Ve=7m/s. ● Q=7.7m3/h;Ve=5m/s

4-3 Analytical expression for collection efficiency

We consider that all the flow rate Q flows through Region II. The variation dC of the concentration of the light dispersion between two cross sections separated by dx can be written:

(5) $Q.dC = -W.\pi.RO.C.dx$

W is the migration velocity of the droplet at the boundary between region II and region I (r = RO/2) and can be written according to Stokes law:

$$(6) \quad W = \frac{\Delta\delta.a.d^2}{18.b.\mu}$$

Δδ being the density difference between the water and the suspension, μ the viscosity of water, d the diameter of the droplets and b a correction factor depending of the size of the droplet. We can take, according to Ref.4:

$$b = 1 + 0,11.Re^{.687}$$

Re being the Reynolds number of the droplet.

According to section 3, we can calculate the acceleration "a" in a given cross section from the mean value of the velocity between the wall and the boundary r = RO/2,that is:

(7) $a(x) = \left[VO(x)+V \right]^2 / 2RO$

We know from section 3 that V(r=RO/2) is constant all along the separation chamber and can be written V = 1.6 Vs, which leads, according to Eq (4),to:

(8) $V = 1.6 \; VO / (1+K.L)$

whereas

(9) $VO(x) = VO /(1+K.x)$

Combination of Eq. (5) to (9) gives the equation governing the variation of the light dispersion concentration along the separation chamber, which is:

$$Q.dC/C = - \frac{\Delta\delta.\pi.d^2.VO^2}{36.b.\mu} \left[\frac{1.6}{1 + K.L} + \frac{1}{1 + K.x} \right]^2 . dx$$

This equation can be integrated for x = 0 to x = L to give the total variation of the concentration of droplets of diameter d between the inlet of the separation chamber (Ci) and the outlet (Co):

$$Co/Ci = \exp\left[-\frac{\Delta\delta.\pi.d^2.V0^2}{36.b.\mu.Q}(L1 + L2 + L3)\right]$$

with:

$$L1 = \left[1.6/(1+KL)\right]^2 . L$$

$$L2 = L / (1+KL)$$

$$L3 = 3.2 \log (1 + KL) / k.(1+KL)$$

The collection efficiency of the cyclone for droplets of diameter d is by definition E = 1 - Co/Ci

Fig.7 gives an example of the efficiency curves calculated with the above expression for the cyclone for the three different flow rates corresponding to the experiments.

5 SEPARATION TESTS RESULTS

The theoretical curves of Fig.7 were used to predict the overall collection efficiency of the hydrocyclone from the experimental drop size spectra. These predictions could be compared with measurements

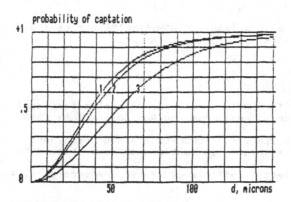

Fig.7 Calculated collection efficiency curves
Oil density = 840Kg/m3; R0=.03m; L=0.6m
1: Q=7.7m3/h;Ve=5m/s. 2: Q=6.6m3/h;Ve=4.3m/s; 3: Q=3.3m3/h;Ve=2.1m/s

on oil in water dispersions. Experimental efficiency curves could not be obtained since drop breakup occuring at the outlet orifice modified the spectra obtained at the end of the separation chamber. The operating conditions, experimental and predicted efficiencies are given on Table I for which the mean inlet drop diameter is defined by:

$$\bar{d} = \frac{\Sigma\ ni.di^3}{\Sigma\ ni.di^2} \quad,\ ni\ being\ the\ number\ of\ droplet\ of\ diameter\ di.$$

It can be seen that the agreement is good, which seems to validate the simple modelling proposed.

Q m3/h	q m3/h	d microns	Ve m/s	E exp. %	E calc. %
3.3	.16	20	2.1	26	26
3.3	.16	33	2.1	40	45
3.3	.16	50	2.1	59	63
6.6	.19	33	4.3	55	62
6.6	.19	47	4.3	66	74
6.6	.19	70	4.3	80	86
7.7	.23	37	5.0	61	63
7.7	.23	49	5.0	73	78
7.7	.23	64	5.0	80	84

TABLE I : Separation Tests Results
Fresh Water, Temp. 25°C, Oil density 840 Kg/m3

6 CONCLUSION

The proposed modelling gives a way to predict approximately the collection efficiency of hydrocyclones treating light dispersions. This model can be used either for the design or for the scaling up of liquid/liquid separators. The validity of the model was demonstrated with a cylindrical hydrocyclone, but if the flow field is known, the method could be extended to conical hydrocyclones.

ACKNOWLEGEMENTS

Thanks are due to TOTAL and ALSTHOM NEYRTEC for having partially supported this work.

REFERENCES

1 COLEMAN.D.A - THEW.M.T Correlation of separation results from light dispersion hydrocyclones. Chem.Eng.Res.Des. Vol.61 1983

2 WOILLEZ.J - LECOFFRE.Y - SCHUMMER.P A high efficiency liquid/liquid separator. 4th Int. Conf. on Multiphase Flows. Nice BHRA 1989

3 DIETZ.P.W Collection efficiency of cyclone separator. AIChe J Vol27 No 6 1981

4 CLIFT - GRACE - WEBER Bubbles, Drops and Particles A.P 1978

Chapter 17

UNDERSTANDING THE FOULING PHENOMENON IN CROSS-FLOW MICROFILTRATION PROCESSES

P Schmitz, C Gouveneur and D Houi (Institut de Mécanique des Fluides de Toulouse, France)

ABSTRACT

Using hollow fiber modules for the production of drinking water needs to investigate laminar flow of a suspension inside a porous tube with wall suction.
At first, numerical and experimental models, valid for similar Reynolds number, are developed to determine velocity and pressure fields at a macroscopic fiber scale for a Newtonian fluid without particles. In order to have a realistic prediction for optimisation of the working conditions of the industrial process we must incorporate time dependant wall permeation characteristics due to surface deposit formation and clogging during the filtration period.
A local 2D flow approach at a microscopic pore scale reveals conditions for particle deposition on the filter surface.
In the second part we propose a statistical model, able to predict the aggregation of micronic particles. Moving and sticking rules are developed taking into account not only the hydrodynamical conditions but also phenomena such as physicochemical, double layer, brownian, concentration, density and particle shape effects. The influence of capture mechanisms on the structure of surface deposits is analysed and characterized by 2D macroscopic mean porosity and thickness quantities.

NOMENCLATURE

α : incidence angle
θ_i : incident sticking angle
θ_v : vertical sticking angle
ν : fluid kinematic viscosity
D_i : porous tube inside diameter
e : thickness of porous media
K : permeability of porous media
L : porous tube length
N_r : number of reentrainment
N_p : number of aggregated particles
P : filter porosity
R_d : filter pore diameter over particle diameter ratio
Re : axial Reynolds number based on U_0
Re_v : radial Reynolds number based on Vw_0
Rms: root mean square
Yc_i : distance from the center of particle i to the wall

1. INTRODUCTION

Recently, membrane separation processes such as cross-flow ultrafiltration and microfiltration are finding applications in a variety of new laboratory and industrial processes. However, they all suffer from "fouling", a limiting and potentially serious phenomenon leading to flux decline during the filtration operation.

Water Treatment – Proceedings of the 1st International Conference, pp. 173–180

Using hollow fiber modules for drinking water production is of high interest because of two main advantages which are : absence or reduction of chemical treatments and compactness of production plants. But the technological process and especially the flow conditions need to be severely optimized because of the low market price of drinking water.

In the 70s, a great number of models have been developed to describe the laminar flow properties in porous tubes. The similarity or fully developed flow analyses, first investigated by Berman (1953) in channels and Yuan et al (1956) in tubes, has been recently used by Chatterjee et al (1986) to model the hydrodynamical field in a spiral wound membrane module assuming uniform wall permeation. More general solutions called the entrance region or developing flow region were proposed (see for instance Gupta et al, 1976) and the existence of regions where the similarity assumption is not valid has been proved, also confirmed mathematically by Brady (1984). Gill et al (1973) optimise the design of hollow fibers for reverse osmosis considering the suction velocity as a function of radial pressure drop across the wall. Gallowin et al (1974) use the permeability as a fitting parameter to find good agreements with experimental measurements in a dead-end porous tube performed by Quaile and Levy (1973), still being the only data for theoretical model validation. Inlet flow and wall filtration conditions in hollow fibers need to consider the developing flow assumption and a nonuniform suction velocity along the tube to calculate the hydrodynamical field. Indeed, the boundary condition at the wall is modified during water filtration because of negative effect due to particle deposition.

Local studies at microscopic scale of Spielman et al (1970), Overbeek (1984), De Gennes (1981) describe the hydrodynamical and physicochemical phenomena ruling the behaviour of suspended particles in the wall region. Belfort et al (1985) apply Cox and Brenner method (1968) to study the influence of inertial and suction effects on particle trajectory. Those interesting works which consider only one or a few particles cannot be extended to a multitude of particles in order to modelize and to understand the formation of surface deposits at the porous wall of a fiber.

The purpose of this paper is to understand the fouling phenomenon in crossflow microfiltration processes by modelling the behaviour of a suspension flowing parallel to a filter surface, like a membrane.

At first we develop a finite element model to compute the velocity and pressure fields inside hollow fibers valid for microfiltration of a Newtonian fluid without particles. Those results are compared with experimental measurements of pressure variations in a porous tube. In order to have a realistic predictive model for the industrial process, we need to understand and predetermine the time dependant wall filtration characteristics due to surface deposit formation and clogging.

Subsequently, the influence of phenomena occuring in the proximity of the wall such as hydrodynamical, physicochemical, double layer, brownian, concentration, density and particles shape effects are taken into account in a statistical model which is able to predict the aggregation of micronic particles on a porous wall. Moving and sticking rules are employed to characterize the respective importance of those different effects on the behaviour of particles launched individually from a random initial location. Such theoretical deposits are qualitatively in good agreement with deposits experimantally observed.

The final objective will be to predetermine the structure and the physical parameters, porosity and thickness, of a deposit for a given suspension, allowing for fiber scale flow computation with more realistic wall boundary conditions.

2. FLOW IN A POROUS TUBE AT MACROSCOPIC SCALE

2.1 Models
Numerical and experimental models are developed simultaneously in order to predict the hydrodynamical field and pressure drop for incompressible laminar Newtonian flow in a porous tube with varying wall suction.

A finite element technique applying a penalty method is employed to solve numerically the axisymmetric Navier-Stokes equations for a wide range of axial Re and radial Re_w Reynolds numbers and wall permeabilities. This study is limited to a parabolic inlet profile. The dimensionless form of the governing equations reveals similarity parameters: a longitudinal Reynolds number $Re = U_0 . D_i/\nu$, a radial Reynolds number $Re_w = Vw_0 . D_i/\nu$, and a shape factor $K^* = K/e . D_i$.

An experimental set-up has been developed (Fig. 1) using a porous tube instead of a microfiber. Working conditions are similar as the dimensionless numbers are kept equal. The laminar flow of silicon oil is investigated at the entrance and at the end of the porous tube. Pressure drop are measured along and across porous tube. Velocity profiles obtained by Laser Doppler Velocimetry characterize the flow in the cross section of the duct. The axial evolution of these hydrodynamical parameters are obtained by repeating measurments for the same experimental conditions when increasing porous tube length.

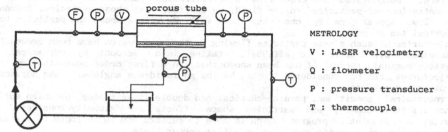

METROLOGY

V : LASER velocimetry

Q : flowmeter

P : pressure transducer

T : thermocouple

Fig. 1 : Experimental setup. Silicon oil flow (ν = 50 cSt) in a ceramic porous tube (L = .5 m, D_i = 3.10^{-2} m, e = 10^{-2} m, K = 10^{-12} m²) with wall suction

2.2 Results

The bulk numerical model is applied in order to predict the evolution of the hydrodynamical field for fixed crossflow filtration conditions.

Computed pressure variations for a porous tube are compared to experimental data obtained for increasing filtration flux (Fig. 2). The figure shows good agreement between computed and measured values. Numerical results have been corrected in order to take into account the over pressure drop due to porous wall of the tube when Re_w=0.

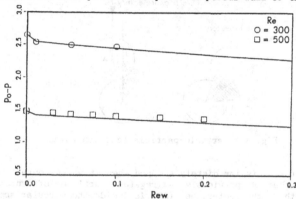

Fig. 2 : Evolution of the dimensionless pressure as a function of Re_w

The numerical work has been developed for the special case of flow in a dead ended porous tube with wall suction (SCHMITZ 1989). Computed pressure variations are compared to experimental data of Quaile and Levy to validate the model. Our model was performed to predict axial pressure variations for backwashing in membrane separation processes using hollow fibers with an external membrane skin.

2.3 Discussion

In the real case of water crossflow filtration we have to take into account a dilute suspension of particles instead of a single fluid phase. The experience proves fouling of membranes during the filtration process leading to a decrease yeld. A study of the flow at microscopic scale performed by particle trajectories analysis shows conditions under which particles deposit on the porous wall and lead to the formation of a fouling cake (SCHMITZ, GOUVERNEUR, HOUI 1989).

In order to take into account the changing boundary conditions at the interface between fluid and porous media due to the formation of a cake, it seems very important to characterize the structure of the deposit.Parameters such as the permeability and the thickness of the cake are obtained by modelling the formation of a cake consisting of spherical particles.

3. SIMULATION OF DEPOSIT FORMATION

In membrane separation process concerning water purification, surface deposits we can indentify consist generally of solid micronic particles, colloidal objects and macromolecules. In this preliminary approach, we are especially interested in the aggregation of a large number of solid particles from a suspension on a filter surface, encountered in cross-flow ultrafiltration conditions. Assuming a low concentration of particles in the fluid, we consider that those particles, driven by the flow, appear one by one near the wall. We will suppose the particles to be spherical and monodispersed.

Trajectories of such single particles flowing in the wall region have been precedently determined by Schmitz et al(1989), taking into account hydrodynamical and physico-chemical forces. It has been shown that, in a first order approximation, such trajectories can be considered linear, having an incidence angle with the horizontal porous plane surface.

Hydrodynamical conditions, physico-chemical and double layer forces, brownian motion, concentration, density and particles shape effects are globally considered in a statistical simulation program which is able to perform and furthermore to visualize the deposition of micronic particles on a flat porous plate.

3.1 Statistical Model

In a rectangular 2D domain, we inject solid circular particles, one by one, following linear trajectories from a random initial location far from the wall. They move towards the filter surface represented by a flat plate with rectangular holes. Special moving and sticking rules are proposed to define the behaviour of particles in the flow and when they make contact with the filter or with another particle already sticking to the wall. Those empirical rules are given to characterize the respective importance of the different phenomena which occur in the filtration zone.

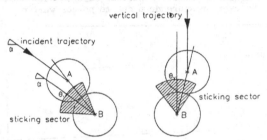

Fig. 3 : Particle-particle adhesion rules

Adhesion rules

A particle of center A is immediately stopped when it makes contact with the filter wall. When it touches a previously aggregated particle of center B, it is also captured provided that the centerline (AB) is inside the circular sector defined by the sticking angle θ as shown in Fig. 3. We then obtain two parameters θ_i and θ_v which determine respectively adhesion in incident and vertical directions.

Displacement rules

A linear incident way is given to each particle A launched from its initial point, characterized by the incidence angle α, another parameter of the model. After making contact with a particle of center B, it turns left or right around this particle until reaching a vertical or a new incident trajectory and so on. The type of this new direction followed depends on the number of reentrainment permitted. The number of reentrainments is a parameter which can be choosen freely and accounts for the maximum number of possible "contacts" before a particle is finally captured. The different cases which occurs are detailed in Fig. 4.

Filter surface

The geometrical characteristics of the filter are taken into account by the porosity parameter P and by the diameter ratio R_d defining the pore size over the particle size. Of course, particles are not always captured as they can pass the filter if their diameter is smaller than the pore diameter.

A realistic deposit on a porous surface is obtained when a multitude of particles move one by one from their initial location to the dynamical collector composed of the filter and the previously captured particles. In this dynamical simulation, the trajectory incidence represents suction intensity over parallel flow. The magnitudes of the two sticking angles θ_i and θ_v characterize the respective importance of perpendicular attractive forces compared to parallel hydrodynamical and perpendicular

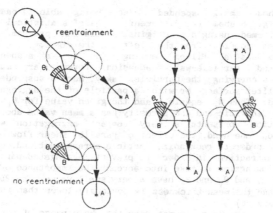

Fig. 4 : Particle displacement rules

repulsive forces. The number of reentrainments enhances the influence of the tangential flow on the behaviour of suspended particles close to the edge and inside the deposit.

3.2 Results

Particle displacement and adhesion rules, such as those previously defined, can be used more generally to simulate the formation of aggregates which have very different structures allowing the simulation of many different applications. Our concern is the understanding of fouling phenomena occuring during drinking water ultrafiltration. Numerical simulations of deposits are carried out for a great number of particles entering individually the specified filtration region, it means physically that such aggregates can be compared to real deposits we can observe after a long water cross-flow microfiltration period. The effect of tangential shear flow on the structure of a solid particle deposit on the porous wall is taken account for in form of a compacity factor which prevents dendrits from being developed by interface erosion and also reduces vertical increase of aggregates before complete space filling. This phenomenon has been experimentally visualized by HOUI (1989) in a cross-flow microfiltration micromodel with video recording of an optical microscop image. Now we specially focus on analysing capture mechanisms by precising their influence on three morphological deposits properties : porosity, density and mean thickness. The final objective of these investigations is to determine the two fundamental macroscopic parameters, permeability and thickness, necessar for the description of the working conditions of a hollow fiber module.

All simulations are developed for a filter surface considered as a nuclepore membrane of 20 % porosity and a pore diameter of 1 µm. The size of solid particles is

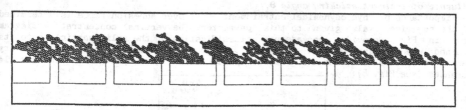

Fig. 5 : Simulation with $\alpha = 20°$, $\theta_i = 60°$, $\theta_v = 1°$, $N_r = 0$

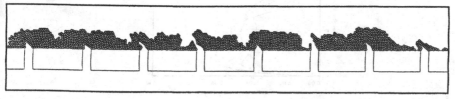

Fig. 6 : Simulation with $\alpha = 20°$, $\theta_i = 1°$, $\theta_v = 1°$, $N_r = 4$

equal to 0.2 µm, those small suspended objects being able to pass the filter pore. Fig. 5 and Fig. 6 show two different types of statistical deposits of 2000 solid particles performed using our original simulation program. In each case, the suspension is assumed to flow from the left to the right.

The first case models a filter fouling due to a suspension flow where particle-particle and particle-wall adhesion forces are much greater than hydrodynamical forces carrying the particles away. Here a suspended particle making contact with the filter surface or with a particle of the aggregate is immediately stopped and captured. Thus N_r equals 0 and the given value for θ_i is about 60°. The deposit structure is dendritic and the porosity has a mean value about 51 %.

The second simulation presents a very compact aggregation we could observe experimentally if, on one hand, the effect of parallel shear flow was efficient even inside the deposit in order to reorganize particle arrangement and, on the other hand, the wall was very attractive in order to prevent aggregated particles from being carried into the flow again. To take into account the importance of tangential flow, N_r is chosen equal to 3, and θ_i and θ_v have a very small value. The 2D mean porosity is then equal to 29 % and the mean thickness is obviously lower than the one obtained in the first case (see Fig. 6).

It now appears to be necessary to investigate the whole range of possible deposits by varying the most important model parameters.

Influence of incidence angle α
This parameter is used to specify the relative magnitude of the parallel shear flow over the radial suction flow. Radial variations of concentration are plotted in Fig. 7 for a large range of α. It is seen that aggregates become more and more compact from 10° to 30°, after this point they keep approximatly the same structure. This is also demonstrated by the continuously decreasing porosity shown in Fig. 8.

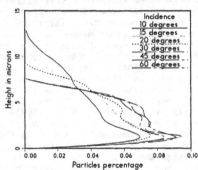
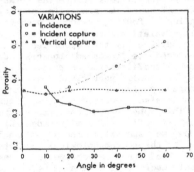

Fig. 7 : Vertical concentration variations
      with θ_i = 1°, θ_v = 1°, N_r = 1

Fig. 8 : 2D mean porosity
      variations

Influence of incident sticking angle θ_i
The importance of hydrodynamical entrainment force over adhesion force is determined by the numerical value given to this parameter. The vertical concentration diagram shown in Fig.9 provides an adequat classification of the solid particle deposits obtained when varying θ_i from 5° to 60°. Of course the small values lead to a very compact structure and the mean porosity increases quickly if the sticking angle increases (see Fig. 8).

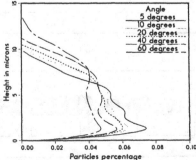

Fig. 9 : Vertical concentration variations with α = 20°, θ_v = 1°, N_r = 0

Influence of the number of reentrainment N_r

This parameter fixes the sensitivity of a suspended particle concerning entrainment by the flow when it makes contact with a previously aggregated object. The simulations drawn in Fig.10 present deposit morphology variations for different values of the number of reentrainment, from branching out (N_r small) to compact structures (N_r high).

Secondary parameters such as θ_v, P and R_d do not have a significant influence on the morphology of theoretical deposits.

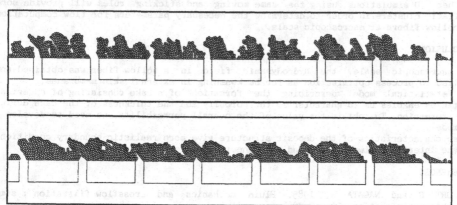

Fig. 10 : Simulation for N_r = 0 (up) and 3 (down) with α = 20°, θ_i = 1°, θ_v = 1°

3.3 Discussion

The main question is : do such statistical aggregates really represent what happens when a filter surface is clogged by solid particle deposition during a cross-flow microfiltration process ? Of course it is very difficult to give an answer because of the difficulty to experimentally observe phenomena appearing at a microscopic scale (the pore diameter is about 1 μm and the particle diameter is supposed to be smaller). Experiments already performed by HOUI (1989) in a 2D micromodel for the filtration of dilute well-muds have shown surface deposits whose structures are qualitatively in good agreement with some of our statistical simulations. We hope that further investigations using the original technique of nuclear magnetic resonance will provide us with useful information on the time dependant of the thickness all along a hollow fiber.

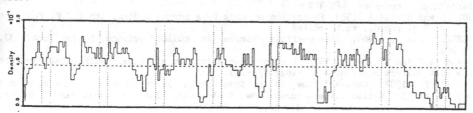

Fig. 11 : Horizontal density variations with α = 20°, θ_i = 1°, θ_v = 1°, N_r = 4

The horizontal density variations, plotted in Fig. 11 for the aggregation depicted in Fig. 6, show that the statistical simulator ensures a uniform distribution of particles all over the studied filter surface except at the downstream and upstream boundaries. Furthermore it can be concluded that deposits formed during a cross flow microfiltration process have a uniform thickness over a filtration length covering a few pores.

It is important to notice that statistical deposits are homogeneous. Bensimon (1983) and Meakins (1985) have demonstrated that this can be proved by the following relation, setting ε equal to 1 :

$$Rms = B \cdot N_p^\varepsilon \quad \text{where} \quad Rms = \sqrt{1/N_p \sum_{i=1}^{N_p} Yc_i^2}$$

It has been well verified for each simulated aggregate. Capture mechanisms do not induce heterogeneity.
Furthermore, the macroscopic quantity Rms can be used to define a mean thickness which takes into account all the roughnesses of the deposit seen at the microscopic scale.
Unfortunatly, it is impossible to calculate the permeability because of the 2D approach, except if we admit the validity of the analogy between 2D and 3D deposits concerning porosity. Then the Brinkman method can be used to calculate the permeability of an aggregate of spherical solid particles.
Further 3D simulations using the same moving and sticking rules will provide more realistic clusters in order to determine the necessary parameters for flow computation in hollow fibers at macroscopic scale.

CONCLUSION

At macroscopic scale, the hydrodynamic field in a hollow fiber was obtained and applied to process optimisation as for backwashing in external skin membranes.
The statistical model describing the formation of a cake consisting of spherical particles enables us to characterize the porosity and the thickness of the cake during its formation. To obtain the value of the overall permeability, we need to develop a 3D model.
These characteristics of the deposit structure give more realistic boundary conditions at the interface between fluid and membrane to compute flow inside hollow fibers.

REFERENCES

BELFORT G and NAGATA N, 1985, Fluid mechanics and crossflow filtration : some thoughts, Desalination, 53, 57-79
BERMAN A.S, 1953, Laminar flow in channels with porous walls, J. Appl. Phys., 24, 1232-1235
BRADY J.F, 1984, Flow development in a porous channel and tube, Phys. Fluids, 27, 1061-1067
CHATTERJEE S.G and BELFORT G, 1986, Fluid flow in an idealized spiral wound membrane module, J. Membrane Sci., 28, 191-208
COX R.G and BRENNER H, 1968, The lateral migration of solid particles in Poiseuille flow -I. Theory, Chem. Eng. Sc;, 23, 147-173
DE GENNES P.G, 1981, Dynamics of concentrated dispersions : a list of problems, Phys. Chem. Hydro., 2, 1, 31-44
GALLOWIN L.S et al, 1974, Investigation of laminar flow in a porous pipe with variable wall suction, AIAA J., 12, 1585-1589
GILL W.N and BANSAL B, 1973, Hollow fiber reverse osmosis systems - Analysis and design, AIChe J., 19, 823-831
GUPTA B.K and LEVY E.K, 1976, Symmetrical laminar channel flow with wall suction, J. Fluids Eng., September, 469-474
HOUI D and RITTER A, 1989, Filtration of muds by a porous medium, 5th I.F.P. Research Conf. on Expl./Prod., to be edited
OVERBEEK J.T.G, 1984, Interparticle forces in colloid science, Powder Tech., 37, 195-208
QUAILE J.P and LEVY E.K, 1973, Pressure variations in an incompressible laminar tube flow with uniform suction, AIAA Paper N72-257
SCHMITZ P, 1989, Laminar flow in a dead ended porous tube with wall suction : application to hollow fibers crossflow filtration, Int. J. Heat Mass Transfer, submitted
SCHMITZ P et al, 1989, Fundamental mechanisms of particle deposition on a porous wall: hydrodynamical aspects, 1st Europ. Conf. on the Math. of Oil Recovery, July, to be edited
SPIELMAN L.A and GOREN S.L, 1970, Capture of small particles by London forces from low speed liquid flows, Env. Sci. Tech., 4, 2, 135-140
YUAN S.W and FINKELSTEIN A.B, 1956, Laminar flow with injection and suction through a porous wall, Trans ASME, 78, 719-724
BENSIMON D, DOMANY E, AHARONY A, 1983, Crossover of fractal dimension in diffusion-limited aggregates, Phys. Rev. Letters, 51, 15, 1394
MEAKIN P, 1985, Accretion processes with linear particle trajectories, J. of Colloid and Interface Sci., 105, 1, 240

Chapter 18

ADSORPTION OF TRIHALOMETHANES ON TO ZEOLITES

S R Chorley, D B Crittenden and S T Kolaczkowski (School of Chemical Engineering, University of Bath, UK)

SUMMARY

Fundamental equilibrium and kinetic adsorption data have been generated for the removal of trihalomethanes (THMs) from water using the hydrophobic pentasil-zeolite known as silicalite. The preliminary data indicates adsorptive capacities which are comparable with those of granular activated carbon (GAC). The chief advantage of hydrophobic zeolites is their high thermal stability in air, up to temperatures of around 1100°C, which should make the process of on-site regeneration relatively straightforward. Hydrophobic pentasil-zeolites are also highly selective towards low molecular weight hydrocarbons such as the THMs, as they have a uniform crystalline structure and a pore size which excludes larger molecules.

For single component adsorption, of chloroform, for example, the equilibria are described by the Freundlich isotherm equation over the range of concentrations investigated, ie from $50\mu g\ \ell^{-1}$ to $1000mg\ \ell^{-1}$. Work on other THMs, although less soluble, also indicate a Freundlich isotherm type.

Further work continues on multicomponent equilibrium data, and on acquisition of batch and column kinetic data. In the absence of external mass transfer limitations, the initial kinetic data, again for chloroform, indicates similar rates of adsorption for GAC and the zeolite.

© 1991 Elsevier Science Publishers Ltd, England
Water Treatment – Proceedings of the 1st International Conference, pp. 181–190

Nomenclature

C_e =   concentration of THM in bulk solution at equilibrium (μg ℓ^{-1})

K_F =   Freundlich isotherm constant (dimensioned by eqn 1)

n   =   Freundlich constant (dimensionless)

q_e =   mass of THM adsorbed per unit mass of adsorbent at equilibrium (g g^{-1})

INTRODUCTION

During the last 10-15 years, there has been a growing awareness of the potential health risks associated with the long term consumption of drinking water contaminated with trace quantities of organic compounds. Greatly improved analytical techniques have permitted the isolation and accurate quantification of these contaminants, present in nanogram to microgram concentrations per litre, so that close to 1000 organic contaminants have now been identified in drinking water for all areas investigated (Ram et al, 1986). These compounds enter the water sources from a variety of origins, including domestic, agricultural and industrial wastes, spillages, seepages, run-offs, and leaching of land deposited wastes, and are not eliminated totally by the standard drinking water treatment methods. It was discovered by Rook (1974) that the chlorination of water for disinfection formed a variety of chlorinated organics from the precursors already present in the source water.

Recent attention has focussed on the formation of trihalomethanes (THMs) in this manner; the four compounds most commonly found are:

(1) chloroform, $CHC\ell_3$

(2) dichlorobromomethane, $CHC\ell_2Br$

(3) dibromochloromethane, $CHC\ell Br_2$

(4) bromoform, $CHBr_3$

Concern over the long term toxicological and carcinogenic effects of these compounds has prompted studies of ways to control their formation. This may be achieved in a variety of ways:

(1) precursor removal prior to chlorination;

(2) use of disinfectants other than chlorine;

(3) controlling chlorination reaction conditions, eg pH, temperature, concentration or reaction time, to minimise production of THMs;

(4) removal of THMs from the treated water.

THM precursors are ill-defined macromolecular structures and their selective removal by adsorption can prove slow and difficult (Ram et al, 1986, p239). Disinfectants other than chlorine may prove less effective and/or more expensive to use, *eg* ozone. Equally, precise control of the chlorination reaction is necessarily complex to cope with wide variations of feed conditions, without over or under dosing (Ram *et al*, 1986, p46).

The use of adsorption as a method of removing THMs from drinking water has a number of advantages over the previous options. It can be applied as a discrete unit operation at the end of the treatment process, allowing all other operations to remain unchanged. The process is generally superior to other removal techniques such as air stripping, distillation or biological treatment, in that it usually has lower capital and operating costs, and is more effective (O'Brien *et al*, 1983, pp337-362). Adsorption using powdered activated carbon (PAC) or granular activated carbon (GAC) is already in widespread use for this application, especially in the USA. The main problem is the cost of regenerating the carbon once it is saturated (O'Brien *et al*, 1983, pp337-362). This can either be done *in situ*, requiring moderately expensive and complex plant, or the carbon is sent away for regeneration, and simply replaced with fresh stock.

The purpose of this research project is to investigate the feasibility of using novel synthetic crystalline pentasil-zeolites such as silicalite (Flanigen and Grose, 1977) and similar materials to adsorb THMs from drinking water. Zeolites are porous crystalline alumino-silicates consisting of an assemblage of SiO_4 and AlO_4 tetrahedra joined by oxygen atoms to form a crystal lattice with pores of molecular dimensions. The pore size is precise and uniform, giving them their highly selective nature. The Si and Al atoms are bound together with oxygen atoms, and in the case of the pentasil-zeolites, having a structure based on a double five ring unit (Figure 1) which, stacked in a variety of ways, form different, but related, structures. The ZSM-5 structure is characterised by a ring of ten tetrahedra and may be prepared in a virtually aluminium-free form known as silicalite.

By varying the ratio of Si to Al from one up to about 100, the nature of the synthesised material is greatly affected. At low Si/Al ratios the zeolites have a high affinity for polar molecules such as water, but at a Si/Al ratio of between eight and ten a transition normally occurs, and a hydrophobic nature is introduced whereby the material preferentially adsorbs organics rather than water. The uniform pore size of silicalite, at around 6 Angstrom units, limits this adsorption to low molecular weight organics only, making the material potentially attractive for the selective removal of THMs from water (Shultz-Sibbel *et al*, 1982). The second advantage of zeolitic adsorbents is their high thermal stability in air up to temperatures of about 1100°C, permitting regeneration by simply heating in air, as compared to the

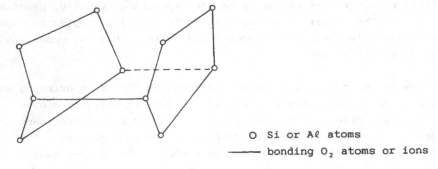

O Si or Aℓ atoms

———— bonding O₂ atoms or ions

Figure 1   Pentasil-zeolite

inert atmosphere and closely controlled conditions necessary for
activated carbon regeneration.

The programme of research will ultimately generate both equilibrium
and kinetic data for a variety of hydrophobic adsorbents in single and
multisorbate systems. This will enable valuable design data to be
developed and comparisons with more established adsorbents to be made.

EXPERIMENTAL METHOD

The fundamental description of an adsorption system is its equilibrium
isotherm. Construction of an isotherm is readily achieved from batch
equilibrium experiments. A known quantity of dried and activated
adsorbent is contacted with a solution of known concentration and
allowed to reach adsorption equilibrium in a closed agitated vessel
maintained at a constant temperature. The solution concentration at
equilibrium is then measured and the process repeated for a range of
concentrations.

Kinetic data may also be obtained by batch experiments. In this case,
a known weight of adsorbent is contacted with a known concentration
and volume of solution in a closed agitated vessel, and the solution
concentration measured at periodic intervals until equilibrium is
reached.

In practice, the following requirements must be fulfilled in order to
obtain accurate and repeatable fundamental data:

(1) The adsorbent must be thoroughly desorbed immediately prior to
    use, to measure true capacity.
(2) The water must be absolutely free of organic chemicals prior to
    preparation of standard solutions of known concentration of each
    THM or mixture of THMs.
(3) Owing to the volatility of THMs, adequate precautions must be

taken to minimise or eliminate losses by evaporation at all
stages, and to use controls to detect and account for any such
losses.

(4) Owing to the low concentrations of THM, accurate and repeatable
analytical techniques are essential for measurement of solution
concentrations.

PROCEDURE

Organic-free water is prepared by distillation and then passed through
an activated carbon filter. Purity is verified by gas chromatographic
analysis. Standard solutions are then prepared by addition of
accurate volumes of THM by micropipette, with subsequent dilution for
lower concentrations in the $\mu g\ \ell^{-1}$ range. The adsorbent is prepared
by heating in a closed porcelain crucible for 4 hours at 400°C
immediately prior to each experimental run. A measured weight of
adsorbent is then contacted with a known volume of THM solution of
known concentration in a screw-topped glass bottle. The bottle is
then agitated in a constant temperature shaker bath for 18 hours, to
attain equilibrium. Similar bottles with different concentrations of
THM are also shaken, along with control bottles with no adsorbent.
After equilibrium is reached, the concentration of the THM is measured
using a Perkin Elmer 8410 gas chromatograph fitted with a 30m x 0.25mm
J & W fused silica capillary column (DB-5 stationary phase), an
electron capture detector and an HS-6 headspace analyser. The
chromatograph is calibrated with standard solutions of the relevant
THM.

From a simple mass balance, the quantity of THM taken up by the
adsorbent from the solution can be calculated for each equilibrium
concentration.

For the kinetic work, a number of identical bottles containing an
equal volume and concentration of THM solution in contact with a known
weight of adsorbent are shaken for 24 hours. Samples are taken from
different bottles at progressive times over the 24 hour period and the
concentration of THM measured as before. Control bottles with no
adsorbent are also shaken and analysed. All equilibrium and kinetic
experiments are carried out at a water bath temperature of 20°C.

RESULTS

A typical isotherm for a silicalite/chloroform/water system is shown
in Figure 2, for the 0-1000 mg/ℓ equilibrium concentration range. The
data obtained is fitted well by the standard Freundlich isotherm
equation:

$$q_e = K_F\ C_e^{\frac{1}{n}} \tag{1}$$

where q_e =   mass of THM adsorbed per unit mass of adsorbent at
                   equilibrium (g g^{-1})

        C_e =   concentration of THM in the bulk solution at equilibrium
                   (μg ℓ^{-1})

        K_F =   Freundlich constant

        $\frac{1}{n}$ =   Freundlich constant (dimensionless)

Taking logs of equation (1) yields:

$$\log q_e = \log K_F + \frac{1}{n} \log C_e \qquad (2)$$

Therefore a plot of log q_e *versus* log C_e over a range of concentrations
should yield a straight line whose intercept and slope give values for
K_F and $\frac{1}{n}$ respectively.   Such a plot is given in Figure 3 for the

Fig 2   Isotherm
CHCL3/Silicalite System

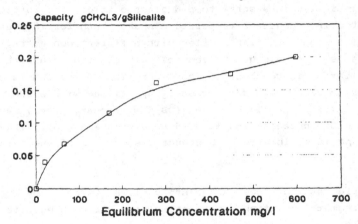

Fig 3   Freundlich Plot
CHCL3/Silicalite System

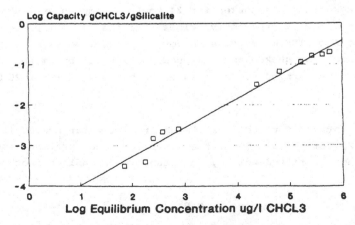

silicalite/chloroform/water system over a broad equilibrium concentration range covering $\mu g \ \ell^{-1}$ to mg ℓ^{-1} THM concentrations.

Values for the Freundlich constants K_F and $\frac{1}{n}$, calculated from these results, are also given in Table 2.

A similar plot is given in Figure 4 for a carbon/chloroform/water system over the same range, for comparison. The granular activated carbon used was Filtrasorb 400, supplied by Chemviron Carbon, and was used as supplied.

Figure 5 shows a comparison of THM concentration *versus* time curves for a silicalite/chloroform/water system and a carbon/chloroform/water system. The same weight of adsorbent was used in each case, *ie* 5mg.

Fig 4   Freundlich Plot
CHCL3/GAC F-400 System

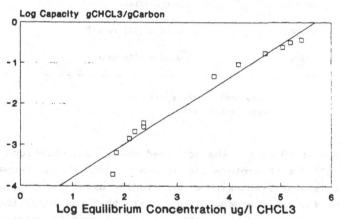

Fig 5   Uptake Curve.
Silicalite and Activated Carbon

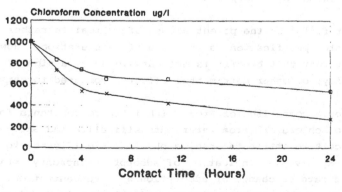

DISCUSSION

The results clearly demonstrate that silicalite is capable of
adsorbing chloroform from aqueous solution over a broad concentration
range and research currently in progress indicates that all THMs can
be adsorbed in this way. Initial experiments with dibromochloromethane, $CHC\ell Br_2$, indicate that the brominated THMs are more strongly
adsorbed than chloroform. For example, for a liquid phase

Table 1   THM solubilities in water at 25°C

THM	Solubility in water
$CHC\ell_3$	0.82 g/100g water[a]
$CHC\ell_2Br$	not available
$CHC\ell Br_2$	0.02 g/100g water[b]
$CHBr_3$	0.32 g/100g water[a]

[a] Lange (ed), 1985, pp214, 650
[b] Experimental data

concentration of 60 mg ℓ^{-1} the adsorbed phase concentrations are 0.072
and 0.10 g g^{-1} for chloroform and dibromochloromethane, respectively.
This result is consistent with other studies using carbon (Faust et
al, 1987, p214), and the general principle that the lower the
solubility of a compound (see Table 1) the greater its extent of
adsorption. This is because the bonding between a compound and the
water in which it is dissolved must be broken before adsorption can
occur. Hence the lower the solubility, the weaker the bond, and the
greater the extent of adsorption.

An important factor in the potential use of silicalite rather than
carbon in water purification is its ease of regeneration. However, it
is important that this benefit is not outweighed by a significantly
lower capacity, or other detrimental effects, compared to carbon.

Table 2 gives a comparison of Freundlich isotherm constants for the
adsorption of chloroform from water onto silicalite and GAC. Whilst
it is difficult to attribute precise physical significance to the
parameters, K_F gives an indication of adsorption capacity, whilst $\frac{1}{n}$ is
a measure of rate of change of adsorption with concentration. As
shown, the constants are very similar indicating little difference in

Table 2   Freundlich isotherm constants for
           silicalite/chloroform/water and
           GAC/chloroform/water

adsorbent	K_F (defined in eqn 1)	$\frac{1}{n}$
silicalite	1.41×10^{-5}	0.75
GAC[a]	1.65×10^{-5}	0.60
GAC[b]	1.58×10^{-5}	0.85

a Faust et al, 1987, p214
b this paper, Figure 4

the adsorptive performance of silicalite and GAC. The kinetic data in Figure 5 indicates a somewhat more rapid uptake of THM by carbon than by silicalite. Whilst the same weight of adsorbent and the same degree of turbulence were used in each case, differences in particle density resulted in fewer particles of silicalite than carbon being used. Other experiments carried out using silicalite alone show the pronounced influence of particle size on rate of uptake. These experiments examined the rate-controlling steps in the adsorption process, to determine the extent of control by internal or external mass transfer. Varying the degree of liquid turbulence during shaking produced only a marginal difference in rate of uptake, whereas crushing and sieving the silicalite to a smaller particle size greatly increased the rate of uptake. Much further work is to be carried out on multicomponent equilibrium studies, and batch and column experiments to determine kinetic data. These initial results, however, indicate an adsorption performance for silicalite broadly comparable to GAC.

The economics of industrial applications cannot currently be evaluated as bulk manufacturing costs for silicalite are not available. However, the manufacturing process for silicalite and similar silica-based polymorphs is relatively simple (Flanigen and Grose, 1977), and production costs should be comparable with many other zeolitic adsorbents which find widespread application in the process industries. Ease of regeneration for silicalite should also yield reduced operating costs in an industrial situation.

CONCLUSIONS

Fundamental data so far obtained indicates that silicalite, a crystalline zeolitic adsorbent, should be capable of effectively removing trace chloro-organic compounds, in particular trihalomethanes, from water. The equilibrium isotherm for the adsorption of $CHCl_3$ onto silicalite is comparable with adsorption onto GAC, with the considerable advantage for silicalite of simple regeneration by heating in air. Equilibrium data can be described over a broad range of THM concentrations by the Freundlich isotherm. The constants in the isotherm are comparable with those for GAC.

Rates of adsorption of chloroform from water by silicalite are comparable with those of GAC.

ACKNOWLEDGEMENTS

This project is sponsored by the UK Science and Engineering Research Council under its Separation Processes Initiative and has the supporting interest of the Water Research Centre, Stevenage. Also, the supply of Filtrasorb-400 activated carbon by Chemviron Carbon is gratefully acknowledged. The authors are also particularly grateful to the staff of Wessex Water for their helpful advice in establishing the analytical procedure used in this research programme.

REFERENCES

Faust, S D and Aly, O M (1987), Adsorption Processes for Water Treatment, Butterworths, Boston

Flanigen, E M and Grose, R W (1977), US Patent 4,061,724, December 6 1977

Lange, N A (ed) (1985), Lange's Handbook of Chemistry, 13th Edn, McGraw-Hill, New York

O'Brien, R P, Rizzo, J L, Schuliger, W G (1983), Removal of Organic Compounds from Industrial Wastewaters Using Granular Carbon Column Processes, in B B Berger (ed) Control of Organic Substances in Water and Wastewater, EPA-600/8-83-011, pp337-362

Ram, N M, Calabrese, E J, Christman, R F (1986), Organic Carcinogens in Drinking Water, Wiley, New York, p4

Rook, J J (1974), Water Treat Exam, 23, 234

Schultz-Sibbel, G M W, Gjerde, D T, Chriswell, C D and Fritz, J S (1982), Talanta, 29, 447-452

Chapter 19

DEVELOPMENTS IN ION EXCHANGE DENITRIFICATION: THE 'WASH-OUT' PROCESS

G S Solt (Cranfield Institute of Technology, UK), I J Fletcher (Hydrotechnica Ltd, Shewsbury, UK)

SUMMARY

Ion exchange is at present the only denitrification process being installed in Britain. There is much room for further improvement of the process, which suffers from several disadvantages. As development of ion exchange by direct experimentation is extremely tedious, a computer simulation has been developed specifically for this problem. It has led the way to an improved process for denitrifying high-sulphate waters using conventional Type II anion resin, which appears to compete favourably with the use of nitrate-selective resin. The process depends on the rapid change in the resin's affinity for polyvalent ions with changing concentration in the liquid. Further development continues.

1. HISTORICAL BACKGROUND

In recent decades nitrate levels in groundwater sources have been rising steadily in many areas, largely as the result of modern agricultural methods, though there has been fierce dispute just how this comes about. In Britain these disputes have for long obscured the essential issue alternative methods of lowering nitrate levels in order to meet the EEC Directive on potable water having been exhausted, many groundwater sources now require denitrification. Whatever changes are made in agriculture, the number of such sources is likely to go on increasing over the next few years at least.

In Britain, all denitrification plants currently being built or designed are based on anion exchange. Ion exchange has the advantage of being a well-tried industrial process raising no novelty in engineering or control. Biological or membrane processes may be used in future, but at present fall short on combined considerations of cost, treated water quality, reliability and ease of operation.

In the ion exchange process water passes through a column of anion exchange resin in the chloride (or chloride/bicarbonate) form. The resin has a higher affinity for nitrate ions and takes them up in exchange for chloride (and bicarbonate). When its capacity is exhausted, the unit is taken out of service and regenerated with brine (or brine and bicarbonate).

2. PROBLEMS OF ANION EXCHANGE DENITRIFICATION

While its engineering is conventional, the process presents some chemical problems when put to this purpose:

a) Conventional anion exchange resin has an affinity for sulphate which is even higher than for nitrate, so that sulphate is also removed, wasting resin capacity and regenerating chemicals.

b) Replacement by chloride, of sulphate, nitrate and some of the bicarbonate in the raw water increases the chloride: bicarbonate ratio in the product, and in many cases results in a corrosive water. Product quality may be restored with a

Water Treatment – Proceedings of the 1st International Conference, pp. 191–198

bicarbonate conditioning following brine regeneration but this adds greatly to the cost and complication of the process.

c)   Due to chromatographic banding in the resin column, the production run at first yields a very high-chloride product, followed by a bicarbonate breakthrough. Such quality variation is undesirable in a mains supply and can only be overcome by several parallel units running out of phase with one another, or very large mixing storage.

d)   When the conventional resin's nitrate capacity is exhausted, the column continues to take up sulphate which displaces a high concentration of nitrate into the product.   It is therefore dangerous to continue the run beyond the nitrate breakthrough point.

e)   The spent regenerant contains the nitrate and other ions removed from the water, plus excess regenerant. Disposal of this solution presents environmental problems whose gravity depends on the location.

All these problems are aggravated by high sulphate in the raw water.   Nitrate selective resins have recently become available, with a higher cost and a lower working capacity and regeneration efficiency, but whose selectivity means that little of this capacity is "waste" in removing sulphate, so that the capacity and efficiency with respect to nitrate are effectively higher when denitrifying high sulphate waters.

The normal practice is to treat only a part of the water, removing nitrate to a very low level, and to blend back with untreated water for a product whose nitrate content is within the EEC directive.   This reduces costs and eases the problems in b) and c) above. Counterflow regeneration is normally necessary to achieve the low nitrate level required to make this practical.

3.   INVESTIGATIONS INTO ION EXCHANGE

Experimentation on ion exchange columns is slow with a column normally taking three cycles of run and regeneration to reach an operating equilibrium. Typically a laboratory test column will produce little more than one experimental result per week's experimentation, whose results are in the form of breakthrough curves.   These reveal nothing directly about events within the column.

In the past ion exchange users have been content with the results obtained from this form of direct experimentation, because in the common industrial softening or deionisation applications, which aim at near-total removal of a class of ions, chromatographic banding within the column appears relatively unimportant. Denitrification, by contrast, is a complex chromatographic process, and needs a fresh approach.

4.   COMPUTER MODELLING

Cranfield Institute of technology started work on anion exchange denitrification in 1984, supported at first by the Science & Engineering Research Council, and currently by the South Staffordshire Waterworks Co. It was identified early on that for reasonable progress a computer model was needed to simulate the process. The Cranfield model is based on the concept of the column as a series of Theoretical Plates (TPs) in which the liquid comes to equilibrium with the resin in each plate before passing on to the next plate.

The equilibrium state in each TP is assumed to have been reached when the four ions under consideration are all in binary equilibrium - i.e. the equilibrium constants for the exchange reactions $Cl-NO_3$, $Cl-SO_4$, and $Cl-HCO_3$ are all satisfied.   Curiously, no resin manufacturer investigates these fundamental properties of their products, and we therefore had to determine them ourselves - a laborious task, made difficult by analytical problems of determining small concentrations of one ion in the presence of large interfering concentrations of another.

Once the model was working, we estimated the height of column equivalent to a theoretical plate (HETP) by trial and error, matching computer runs with actual experimental data, to give us the number of plates (NTP).

After continuing improvement, the simulation can be operated with regeneration in co- or counterflow, with additional regeneration stages if required. The NTP can be changed for different stages of the cycle, to simulate differences in kinetic conditions e.g. between regeneration and run. A leakage factor, by which a preset fraction of the flow "by-passes" each successive plate has improved its accuracy. Any number of successive cycles can be run without interruption, with termination of the production run controlled either by nitrate breakthrough, or to a fixed volume of water treated.

The model is far quicker than direct experimentation. Its speed gives us the freedom to explore even unlikely alternatives, with occasional column tests to check the accuracy of prediction. With conventional resins the model's predictions appear to be quite reliable, and we have had occasions when the computer showed up errors in the experimental work.

Results for the selective resins are not quite so accurate, due probably to their non-ideal behaviour. The predictions nevertheless give a useful qualitative guide, even if they need to be checked rather more thoroughly.   Work continues to improve the model's performance in this respect.

The simulation's printout shows the condition of the resin in each plate within the column at all points of the cycle, which has proved particularly useful, as it gives an insight into the workings of all stages of the process.

Now that the facility to simulate the process quickly and cheaply has been more widely appreciated, other investigators in Britain have taken up the technique.

5.   THE 'WASHOUT' PROCESS

In any real applications, only a part - usually 30 to 50% - of an ion exchanger's total capacity is utilised: this is called its "working capacity".   Increasing levels of brine regeneration increase the resins' working capacity, but with diminishing returns at higher regeneration levels.

Examination of the movement of ions within the column, provided by the computer simulation, showed us early in our investigations that in the regeneration step sulphate comes off the column very readily, while a zone of high nitrate merely tends to move backwards within the column without much of it actually being removed. Most of the working capacity was provided by capacity which has been occupied by sulphate.

Ion exchange resins' affinity for polyvalent ions is well known to fall sharply with rising concentration, but the degree to which this affects the ions' relative regeneration efficiency was unexpected.

We set about using the model to explore means of putting this phenomenon to practical use, which eventually led us to the "Washout" process for denitrifying high-sulphate waters, whose cycle runs as follows:

a)   The production run continues normally until nitrate breakthrough, or until a predetermined volume of water has been treated.

b)   In the "washout" step a sulphate-rich solution displaces nitrate and other ions on the resin to waste, and converts the resin largely to the sulphate form.

c)   In the regeneration which follows, the sulphate-laden resin is regenerated so easily that most or all the resulting eluate consists of sodium sulphate, which is saved.

d)   This eluate is diluted with raw water to provide the high-sulphate washout solution (see b) above) in the next cycle.

The ease with which a sulphate-laden resin can be regenerated results in a better regeneration, which in practice means a higher working capacity, or lower regenerant usage, or both. Resin is delivered by the manufacturers in the all-chloride form. Over the first few cycles in use the process progressively accumulates a recycled stock of sulphate. With low-sulphate waters this stock remains low and the improvement is not large. The greatest improvement in performance which is obtained from the use of the "Washout" process (which has been patented) on moderately high-sulphate waters, using conventional rather than selective resin. For very high-sulphate waters the selective resins would still be more economical.

Figure 1 summarises a number of field trials of conventional denitrification of a raw water whose composition is shown in Table 1. Two different nitrate-selective ion exchange resins, and one conventional Type II gel anion resin were regenerated with NaCl.

FIGURE I   NITRATE CAPACITY

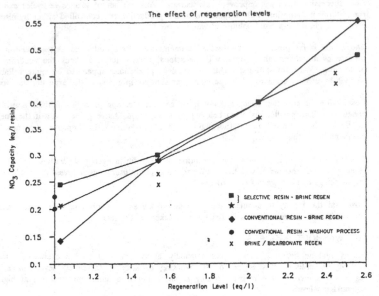

The effect of regeneration levels

The graph shows the resin's working capacity with respect to nitrate only. The conventional resin normally has a low nitrate capacity because it uses up much of its working capacity in removing the sulphate in the water as well. The conventional resin performs relatively badly at the low regeneration level, which is the most economical in terms of regenerant usage.

This water calls for brine and bicarbonate regeneration to give the final blended product the desired chloride:bicarbonate ratio. Two points have been superimposed which show the reduced regeneration efficiency due to lower effectiveness of bicarbonate as a regenerant as compared with chloride. On these points the bicarbonate wash represented 10% and 25% respectively of the total regeneration level.

The graph also shows two "Washout" results for brine/bicarbonate regeneration in which the bicarbonate accounted for 40% of the total regeneration level. Washout was performed with 10 bed volumes of 8 eq/l sulphate solution. Their results suggest that on this raw water analysis the process brings the conventional resin's performance with bicarbonate regeneration to the same level as that of the selective resins, at least at the low regeneration levels which have so far been explored.

This set of process details was determined by an exhaustive computer simulation. Its predictions were tested first on a made-up water in the laboratory and then on the actual water in the field.   The results suggested that the Washout process with co-flow regeneration of conventional resin has a lower operating cost (in terms of regenerant usage) than conventional operation of the nitrate-selective resin.

In this case the eluates from the actual regeneration by chloride and bicarbonate are both completely retained for use as washout in the next cycle. Apart from relatively clean streams such as backwash and rinse, only the washout step eluate has to be disposed of, and this contains mostly nitrate and sulphate.

The results shown in Figure 1 are field results. The field trials varied considerably from run to run (as ion exchange processes tend to do) but when averaged, the results confirm the computer's predictions shown in Table 2.   Table 2 also compares the use of a conventional resin with Washout, with that of a selective resin in the normal manner. The criteria for these tests where that the nitrate level of the blended product should not exceed 0.8 mg equiv 1^{-1}, and that its chloride:bicarbonate ratio should not rise above 0.5.

Figure 2 shows the analyses of blended product water produced by these two processes. The last two lines in each half of Table 2 give the waste volume and the total ions discharged by the respective processes. While the Washout process produces a far greater volume, its total salt content, and especially its chloride content, is much smaller.

FIG 2

ANION EXCHANGE DENITRIFICATION - COMPUTER SIMULATION

FINAL BLENDED PRODUCT QUALITY (1:1 BLEND WITH UNTREATED WATER)

A. CONVENTIONAL RESIN USING WASHOUT AND CO-FLOW REGENERATION

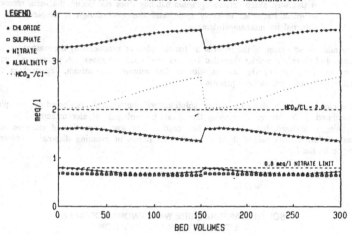

B. NITRATE SPECIFIC RESIN USING COUNTERFLOW REGENERATION

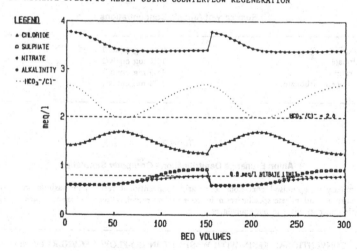

BENEFITS OF THE "WASHOUT" PROCESS

The advantages from the process are more far-reaching than is immediately apparent:

a)   There is a saving in regenerant chemicals when compared with conventional resins. *Because bicarbonate is both a costlier and a less efficient regenerant then chloride, it becomes particularly important in those cases where bicarbonate wash has to be used in order to maintain the treated water quality. In those cases the washout process can even show a saving compared with selective resins.*

b)   The waste produced contains the nitrate and other ions taken out of the water as in the conventional process, but mixed with less surplus regenerant - i.e. less chloride, which is often the limiting factor in effluent disposal.

c)   The improved efficiency of regeneration after washout gives the resin a higher level of regeneration which in turn results in lower nitrate leakage in the product. As a result, the washout process can give good product quality with co-flow regeneration, avoiding the cost and complexity of counterflow plant, its inability to deal with modest levels of suspended solids, and reducing the sharp variation in product quality due to chromatography in the column.

d)   Conventional denitrification plants in the field have been damaged by $CaCO_3$ precipitation, and some prevention is needed. Co-flow plant is unlikely to suffer in this way: sharp chromatographic banding does not occur, the resin receives a complete backwash in every cycle, and the until design is simpler, with less potential for mechanical damage.

The main disadvantage of the process is that the effluent produced by the washout is more dilute and therefore bulkier than that from conventional processes. A promising start has been made to operate the process with smaller volumes of washout, approaching those produced by the conventional process.

It seems likely that dumping nitrate solutions will be severely restricted or completely prohibited in the future. Destroying the nitrate by biological or electrochemical processes is being considered, but is pointless and costly. The low proportion of chloride in the waste solution from the Washout process raises hopes of avoiding disposal problems by re-use on the land.

SOUTH STAFFORDSHIRE WATERWORKS COMPANY - LITTLE HAY PUMPING STATION

Table 1

Summary of Anionic Concentrations

Chloride	0.85 mg equiv/1
Sulphate	1.21 mg equiv/1
Nitrate	1.06 mg equiv/1
Hydrogen carbonate	3.28 mg equiv/1

Table 2

Anion Exchange Denitrification - Computer Simulation

Summary of optimised conditions, comparing conventional resin with washout in co-flow operation, and nitrate-specific resin in counterflow regeneration. (cf also Figure 2). All ionic concentrations in mg equiv/litre.

A. CONVENTIONAL RESIN WITH WASHOUT IN CO-FLOW REGENERATION:

	Cl-	SO_4^{--}	NO_3^-	HCO_3^-	Bed Volumes
Feed	0.85	1.21	1.06	3.28	300
Washout	1.4	32	1.0	1.5	25
Regen 1	1130	0.1	0.1	0.5	0.5
Regen 2	0.1	0.1	0.1	480	0.75
Waste	6.3	11.8	8.1	9.1	25
Total ions in waste	157	295	202	227	= Total 881
(mg equiv per litre resin)					

B. NITRATE SPECIFIC RESIN IN COUNTERFLOW REGENERATION:

	Cl-	SO$_4$--	NO$_3$-	HCO$_3$-	Bed Volumes
Feed	0.85	1.21	1.06	3.28	300
Regen 1	1260	0.1	0.1	0.1	0.75
Regen 2	0.1	0.1	0.1	388	0.75
Waste	168	73	84	50	3.5
Total ions in waste (mg equiv per litre resin)	560	256	294	175	= Total 1285

NB the reduction in total ions to waste, and in particular the reduction in waste chloride.

7.   **CONCLUSION**

A novel procedure for ion exchange denitrification employing a sulphate "Washout" step has been investigated, using computer simulation and tested on a pilot scale plant.

The process is particularly advantageous for higher sulphate waters and offers potential advantages of more economical use of regenerants, improved effluent quality and the use of simpler and cheaper co-flow operation.

Chapter 20

GROUNDWATER POLLUTION BY VOLATILE ORGANIC SOLVENTS: SOURCE IDENTIFICATION AND WATER TREATMENT

M Burley, B D Misstear and R P Ashley (Mott MacDonald Ltd, Cambridge, UK), J Ferry and P Banfield (Anglian Water Services Ltd, Cambridge, UK)

SUMMARY

Groundwater pollution by industrial solvents such as trichloroethene and tetrachloroethene is of increasing concern internationally. These chemicals may break limits set in regulations for drinking water to protect consumer's health. It is necessary in such cases to treat the water to ensure that it is safe to drink. In this paper the authors describe some of their recent experiences in the UK of identifying the sources of contamination and of designing and operating suitable water treatment plants.

One case study has highlighted the necessity of detailed usage surveys of potential contaminants in siting exploration holes for the identification of pollution sources. A second case study has demonstrated the importance of comprehensive monitoring programmes, since the chlorinated solvents were only discovered as a result of another contamination.

Two plants have been designed and constructed using air stripping for the removal of carbon tetrachloride, trichloroethene and tetrachloroethene. The bases for design and operating results are discussed.

NOMENCLATURE AND ABBREVIATIONS

PCE = Perchloroethylene = tetrachloroethene

TCE = Trichloroethene

CTC = Carbon tetrachloride = tetrachloromethane

TCA = 1,1,1 - trichloroethane

VOC = Volatile organic compound

PWS = Public water supply

$\mu g/l$ = micrograms per litre

GC/MS = Gas chromatography/mass spectrometry

x = Mole fraction of the organic in water

z = height of packing (eg ft - m)

L_m = molar liquid flow rate per unit cross sectional area of the tower (eg 1b mole/h/ft² - kgmole/h/m²)

Water Treatment – Proceedings of the 1st International Conference, pp. 199–216

x_2 = mole fraction of solute in inlet water

x_1 = mole fraction of solute in treated water

ρ_m = molal density of liquid (eg 1bmole/ft^3 - kgmole/m^3)

K_L = liquid phase transfer coefficient (eg 1bmole/h/ft^2/(1bmole/ft^3) - kgmole/h/m^2/(kgmole/m^3))

a   = specific surface area of packing eg ft^2/ft^3 - m^2/m^3

D   = diffusivity of the VOC in water (ft^2/h - m^2/h)

L   = liquid rate (1b/h/ft^2 - kg/h/m^2)

μ = liquid viscosity (1b/ft/h - kg/m/h)

ρ = liquid density (1b/ft^3 - kg/m^3)

1.   INTRODUCTION

Volatile chlorinated solvents have been used as non-flammable cleaning and degreasing agents in industry since the development of adequate storage and handling equipment just over 60 years ago. Trichloroethene (TCE) was the most widely used solvent of this group until the introduction of tetrachloroethene (PCE) as a lower volatility solvent after the Second World War. The 1970's and 1980's have seen a decline in use of TCE and PCE in favour of less toxic solvents such as 1,1,1-trichloroethane (TCA) and 1,1,2-trichlorotrifluoroethane (Freon 113), and alternative cleaning methods.

The toxicity of the traditional solvents TCE and PCE in their liquid and concentrated vapour form has been recognised almost since their introduction. Hazards associated with their presence as trace contaminants in drinking water supplies, principally relating to their potential carcinogenic action, were perceived only during the 1970's, and began to have an impact on water supply undertakers and regulating authorities from the middle of that decade (eg, US Environmental Protection Agency, 1975 and European Economic Community, 1976 and 1980).

There was thus a considerable period when industrial cleaning and degreasing processes took little or no account of the potential risks stemming from releases of small quantities of solvents into the aqueous environment. The legacies of the period, in the form of contaminated public water supply (PWS) boreholes, may be revealed only after a long interval and at a distance from the sources of contamination. The discovery of contamination thus presents water supply undertakers and regulators with urgent and unexpected problems, on the one hand relating to identification of sources of contamination, and on the other hand with treating the water so as to bring the source back into supply at the earliest opportunity.

This paper examines two cases in which these problems were addressed, and discusses the solutions that were implemented. Those matters of general applicability where particular difficulties were overcome are described.

2   SOURCE IDENTIFICATION

2.1   Case Study A

A PWS groundwater source in eastern England was found to be contaminated by volatile organic solvents in 1987. The main contaminants detected were TCE (generally at concentrations between 40 and 80 µg/l) and PCE (generally between 5 and 10 µg/l). The World Health Organisation tentative guideline values for these substances in drinking water are 30 µg/l and 10 µg/l respectively (WHO, 1984) and these values have been adopted in recent water supply regulations in the UK (Department of the Environment, 1989).

The PWS source is located on the outskirts of a city near the confluence of two rivers, with a nearby industrial area upstream (Fig 1.) The source comprises two boreholes which abstract from the fissured Upper Chalk aquifer. The chalk in this area is overlain by drift deposits, which typically consist of 5 to 10 m of sands and gravels and glacial Boulder Clay. The upper chalk layers are degraded to a characteristic structureless "putty" texture of low permeability, roughly 10 m thick.
The PWS source was taken out of supply immediately after detection of the contamination and a water treatment system was installed, as discussed in Section 3. Measures were also taken to trace the causes of the contamination. Following a survey of solvent usage, which confirmed that TCE and PCE were widely used within the industrial area, an investigation was carried out to identify the source or sources of pollution. The investigation was performed in two distinct phases:

-       Phase I studies were directed towards determining the nature and extent of pollution in the aquifer;

-       Phase II studies were designed to identify the sources of pollution more precisely.

Methodology

The Phase I investigations included the drilling and test pumping of six exploratory holes, which were subsequently completed as multi-screened piezometers to permit long term monitoring; their locations are shown in Fig. 1. Specific hygiene precautions were taken during drilling to prevent accidental contamination of samples, loss of volatile compounds during storage, or cross contamination between boreholes. For example, all drill tools and casing were flame-cleaned before being used to construct each bore, in order to remove all volatile organic compounds such as degreasing agents, grease and oil.

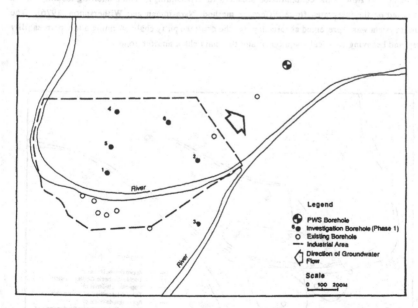

Fig 1 Site Illustration - Case Study 'A' (Schematic)

During the investigation, a large number of samples were collected, including:

-       chalk core samples, analysed for solvents absorbed into the pore structure or dissolved in porewater. Analysis was carried out by disaggregation followed by *steam distillation*, liquid-liquid extraction and gas chromatography, and the results are reported as concentrations in pore water;

-    depth samples of water collected during drilling to determine the vertical chemical profile in groundwater flowing in fissures;

-    bulk samples of water collected during final pumping tests to determine the chemistry of the main groundwater flow zones.

Almost 100 samples were analysed for volatile organic and other compounds, including check analyses on selected duplicate samples.

Phase II commenced with a more detailed survey of past and present solvent usage within the areas of pollution identified during Phase I. The results of this survey were used to target precise locations for further drilling investigations.

Thirteen shallow boreholes were drilled at five locations of known solvent usage, wherever possible within 10 m of a putative source. The aim of the boreholes was to identify the presence of contaminants in the unsaturated zone above the water table, thereby indicating sources of pollution. The drilling, sampling and analytical techniques adopted were similar to those used in Phase I. About 60 drift and putty chalk core samples and water samples were collected and analysed for TCE, PCE, CTC and TCA; about 30 duplicate samples were also analysed.

A groundwater model of the drift and chalk aquifers was set up to investigate the direction and rates of groundwater flow in the contaminated area and its surroundings. The modelling technique used was based on the integrated finite difference method (Narasimhan and Witherspoon, 1976). The aquifer system was represented as three layers: the drift; the putty chalk (forming a low permeability layer, and behaving as a leaky aquitard); and the main chalk aquifer zone.

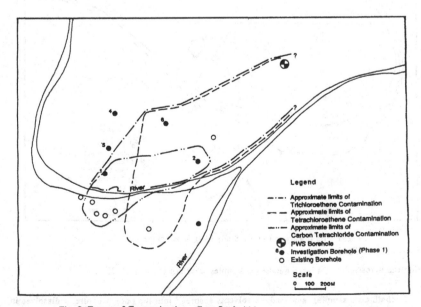

Legend

— · — Approximate limits of
        Trichloroethene Contamination
— — — Approximate limits of
        Tetrachloroethene Contamination
— ·· — Approximate limits of
        Carbon Tetrachloride Contamination
⊕ PWS Borehole
● Investigation Borehole (Phase 1)
○ Existing Borehole

Scale
0   100   200M

Fig 2 Extent of Contamination - Case Study 'A'

Results

The results of the Phase I investigation showed the chalk aquifer to be widely contaminated with TCE, PCE and TCA. Although the concentrations within the main chalk aquifer were generally moderate (10 to 100 µg/l), very high levels of TCE were measured in the upper, low permeability weathered putty chalk at one site (Borehole 6 - 19,850 µg/l) and in the permeable drift above this putty chalk at a second site (Borehole 1 - 25,300 µg/l). Localised pollution by CTC was also detected at two other boreholes. The extent of the main areas of contamination is shown in Fig. 2. The distribution of the pollutants suggested that there have been multiple slugs of organic solvents entering the groundwater system at a number of sources over several years.

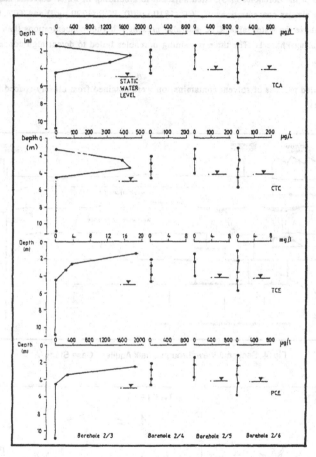

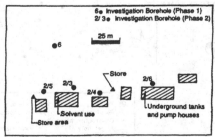

Fig 3 Example of Contaminant Profiles with Depth
       Case Study ' A '

The investigations confirmed the conceptual model whereby chlorinated solvents, spilled at the ground surface, accumulate at the top of the drift, in small depressions on the irregular surface of the low permeability putty chalk. The solvents then seep slowly through this layer into the fissured chalk aquifer below, where flow rates are rapid. This is shown schematically in Fig. 4.

The Phase II shallow drilling programme identified severe contamination by organic solvents at 4 of the 5 locations investigated. The profiles of contamination with depth at one of the sites are shown in Fig 3. All four boreholes at this site were within 100 m of Phase I Borehole 6 that had detected high concentrations of TCE and, to a lesser extent, PCE and TCA at the water table in the chalk aquifer. One of the boreholes (2/3), sited adjacent to a building in which solvents had been used, detected very similar concentrations of TCE (18100 µg/l), PCE (1720 µg/l) and TCA (1740 µg/l), at least 2m above the water table; CTC was also present at higher concentrations than those found at any site during Phase 1. The three remaining boreholes failed to detect high concentrations of contaminants.

Similarly varied profiles of solvent contamination were obtained from closely-spaced boreholes at the other sites.

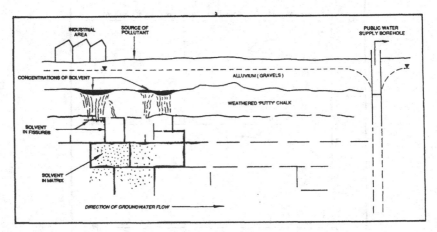

Fig 4 Sectional View through Chalk Aquifer - Case Study 'A'

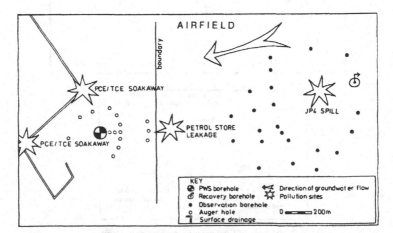

Fig 5 Site Illustration - Case Study 'B' (Schematic)

The groundwater modelling studies were successful in demonstrating the flow pattern within the chalk aquifer and thus the likely flow paths from the sites of known contamination to the PWS source. The modelling studies were unsuccessful, however, in determining the age of pollution. Although the travel time within the chalk aquifer, from the industrial area to the PWS source, was shown to be less than 2 years, a wide range of travel times was estimated in the putty chalk (4 to 30 years), due to uncertainty about the value of effective porosity of this lithology.

Discussion

This study has highlighted the necessity of carrying out detailed surveys of contaminant usage prior to selecting sites for exploratory boreholes. Without such studies, it is unlikely that the sources of contamination would have been successfully identified, because the levels of contamination in the unsaturated zone were found to be extremely localised.

This case history has also highlighted the difficulty in estimating the travel times for contaminant migration, and hence the age of pollution, due to the controlling influence of low permeability layers.

2.2   Case Study B

In the summer of 1978 there was a spillage of a quantity of JP4 aviation fuel at an airfield in eastern England. A PWS borehole source in the chalk aquifer 1.2km distant was threatened. There are drainage ditches in the area but no other significant surface water features. Surface water was drained to shallow soakaways directly into the chalk aquifer. The aquifer is unconfined and has an extremely high transmissivity . The permeability is concentrated in the upper 10 m of the saturated zone, and the groundwater levels respond rapidly to rainfall events. An industrial estate is located upgradient of the PWS source beyond the airfield. Investigations at this site were first reported on by Tester and Harker (1981).

Methodology

Initially, 25 observation boreholes and one oil recovery borehole were drilled on the airfield. Aviation fuel was found on the shallow water table, with a movement of 250 m in the direction of the PWS borehole. A limited amount was recovered but further recovery was impractical due to heavy growth of sulphur bacteria on the borehole.

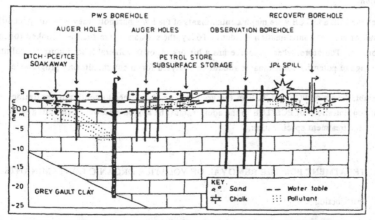

Fig 6 Sectional View through Chalk Aquifer - Case Study 'B' (Schematic)

An additional 14 auger holes were drilled to a depth of up to 10 m, with the objective of providing an early warning of aviation fuel contaminant migration towards the PWS borehole. The shallow depth minimised creation of fast flow paths to the high yielding horizons. One of the auger boreholes was drilled on the side of PWS borehole away from the known contaminated areas, to provide a background control for water quality. Fig. 5 and Fig.6 show the distribution of the auger boreholes and a cross section of the site respectively.

In order to determine the nature, degree and extent of the contamination, a major pump test exercise was carried out on the PWS borehole using the auger holes and boreholes as observation wells, and monitoring the flow in the drainage ditches over a low weir. Geophysical logging was carried out on all available boreholes to identify the major flow horizons in the aquifer. Groundwater quality monitoring was conducted in conjunction with test pumping of the PWS borehole. Samples were collected primarily by depth sampling at the major flow horizons in the aquifer, and analysed by GC/MS for organic constituents.

Results

The JP4 aviation fuel was not found to contaminate the PWS borehole at any stage, despite the quantities in the aquifer nearby.

However, sampling of the auger holes and PWS borehole indicated the presence in the groundwater of PCE and TCE at maximum concentrations of 85 and 20 µg/l respectively. The PWS borehole was removed from supply immediately after discovery of the pollution.

As well as VOCs, other hydrocarbons were also detected in the auger holes. The hydrocarbons were traced to a source at a nearby petrol filling station. The auger holes delineated a "pancake" of petrol of limited areal extent floating on the water table and leading from the petrol station.

The source of the TCE and PCE was never identified but the data suggested that the input of TCE and PCE to the aquifer had occured from one of the drains, in the form of a slug. A sump which drained into the ditch contained TCE and PCE. This was emptied to a recognised waste disposal facility. Depth samples indicated that both total volatile organic compounds and TCE/PCE increased with depth, even though geophysical logging demonstrated clear horizontal layering of flow in the aquifer. In 1985, routine monitoring of the auger boreholes indicated an increase of TCE/PCE.

Discussion

Although the PCE and TCE were major contaminants of the PWS borehole, they were only identified as a result of another contamination incident. Today, the substances are routinely looked for in all PWS sources. Therefore, where a source has a history of, or is vulnerable to, industrial pollution, a wide range of potential contaminants should be analysed for in any monitoring programme.

The variable levels of PCE and TCE in the groundwater, due either to continuing pollution, or to remnant concentrations of PCE/TCE in the aquifer, must be appreciated when predicting worst case scenarios for treatment specification.

3.   TREATMENT FOR THE REMOVAL OF VOLATILE ORGANIC CONTAMINANTS

3.1   Introduction

VOC pollutants can be removed by activated carbon but a much cheaper process, particularly as a first stage treatment is removal by air stripping. The design characteristics for removal are described in Section 3.2. This is followed by a discussion of the operating results for the two water treatment plants installed at the polluted groundwater sources referred to as Case A and Case B. The plant in Case A was designed by Anglian Water and the performance predictions were checked by the Water Research Centre. The plant in Case B was designed by Sir M MacDonald & Partners (now Mott MacDonald) for Anglian Water, following pilot tests by Chemviron Carbons.

3.2   Design Characteristics for Stripping

Absorption and stripping are well established as separation processes in the chemical industry and the design principles are well known.

If the solvent (water) is contacted with air a proportion of the volatile component will leave the water and enter the air. In order to do this efficiently a large air-water interface is required and for this reason stripping is carried out in towers packed with materials which are chemically inert and with a large surface area. Many types of packing have been used : Raschig rings, Pall rings, Intalox saddles packed at random and modular packings are widely used (Green and Moloney, 1984).

The stripping of gases of low solubility is controlled by the rate of diffusion of the solute in the liquid phase and this applies to the VOCs in water.

In the design of a packed tower the following are usually known:

-        the flow rate of liquid to be treated
-        the concentration of pollutant in the raw water
-        the concentration to be achieved after treatment
-        the water temperature

The main parameters to be established are:

-        the height of the packing
-        the diameter of the tower
-        the air flow rate required
-        the pressure drop of air through the tower

The operation of a packed stripping tower is best understood by reference to Fig 7a and Fig 7b.

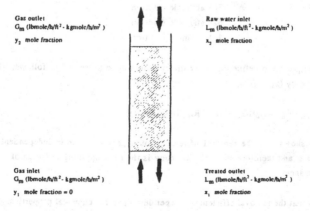

Gas outlet
G_m (lbmole/h/ft^2 - kgmole/h/m^2)

y_2 mole fraction

Raw water inlet
L_m (lbmole/h/ft^2 - kgmole/h/m^2)

x_2 mole fraction

Gas inlet
G_m (lbmole/h/ft^2 - kgmole/h/m^2)

y_1 mole fraction = 0

Treated outlet
L_m (lbmole/h/ft^2 - kgmole/h/m^2)

x_1 mole fraction

Fig. 7a Mass Balance in Tower

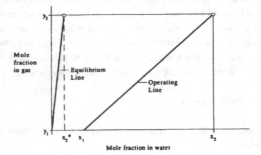

Fig. 7b   Concentration of VOC in Gas and Liquid

In Fig 7a the tower is illustrated with liquid entering at the top and leaving at the bottom. The flow is counter-current with air entering at the bottom and leaving at the top. Subscripts 1 and 2 refer to bottom and top conditions.

The Operating Line

This represents a material balance in the tower

$$L_m (x_2 - x_1) = G_m (y_2 - y_1) \tag{1}$$

The Equilibrium Line

The equilibrium line is a representation of Henry's law which at atmospheric pressure is:

$$x^* = y/H \tag{2}$$

Where x^* is the concentration (mole fraction) of the VOC in the liquid phase which would be in equilibrium when the concentration of the VOC in the vapour is y_1 (mole fraction) and the operating pressure is 1 atmosphere.

Values of H for VOCs are quoted below:

TCE	307 ats/mole fraction
PCE	626 ats/mole fraction
CTC	761 ats/mole fraction

The relationship used to define the tower dimensions can be presented as follows, where R is the removal efficiency $(x_2 - x_1)/x_2$:

$$Z \rho = K_L a (HG_m - L_m) = L_m HG_m \ln \{(1 - RL_m/HG_m)/(1-R)\} \tag{3}$$

Equation (3) shows that the removal efficiency of a packed tower is independent of the inlet concentration x_2 and therefore removal efficiency is the most appropriate means of monitoring the tower performance.

It also shows that the removal efficiency is dependent upon the chemical properties, ρ and H and the design variables Z, L_m and G_m and the mass transfer characteristics $K_L a$.

Specification of Plant Performance

As stated in Section 2 the presence of VOCs in groundwater can be due to contamination which occurred in the past and concentrations are therefore often variable. Consequently, it is usual that the concentrations of contaminants in the raw water at the time of plant commissioning and testing are less than the maximum concentrations to be satisfactorily handled by the plant. Therefore if a performance specification is to be used it is essential to request a specific removal efficiency to guarantee quality of product from a range of raw water concentrations.

In Case B described below both outlet concentrations and removal efficiencies were specified to tenderers as follows:

	Outlet Concentration μg/l	Removal Efficiency %
TCE	12	88
PCE	5	90
CTC	1	80

A further aspect of importance is the potential for the formation of calcium carbonate scale on the tower packing following carbon dioxide removal. If uncontrolled the packing can become completely clogged and cease to function. In order to avoid this problem a scale inhibitor such as Calgon must be dosed to the raw water. The dose rate found satisfactory for such chemicals is 0.3 mg/l as P or 1 mg/l as Calgon chips.

Lagging of pipework has been found necessary for external installations in eastern England if problems of freezing are to be avoided.

3.3   Operating Results from Case Studies

Case studies for two works are presented. In Case A the stripping system has operated very much as expected but in Case B the performance has been somewhat erratic. The causes of this erratic performance have not yet been fully identified. Possible explanations are discussed.

Design Parameters

The design parameters for the two cases are presented in Table 1.

The Performance of Towers

The performance of the towers is presented in graphical form for TCE. Similar trends were observed for PCE. For CTC concentrations in the raw water were so low that concentrations in the aerated water were below detection limits.

The graphical results are presented in two forms:

-    concentrations in the raw and aerated water against time
-    removal efficiency against time

TABLE 1

Design parameters for Stripping Volumes

		Case A	Case B
Overall height m		7.1	7.58
Packed height m		3.22	5.25
Internal diameter m		2.22	1.8
Tower Packing:	Type	Pall rings	Snowflake - 2
	Size mm	22	94
	Material	Polypropylene	Polypropylene

For two towers:

Total liquid flow m³/h	500	500
Total air flow m³/h	13,780	10,000
Cross sectional area m²	7.74	5.1
L_m kgmole/m²/h	3,590	5,363
G_m kgmole/m²/h	77	84.3
L_m/G_m	46.6	63.6
Volumetric ratio G/M m³/m³	27.6	20

3.4   Case Study A

The performance of the Case A columns is shown in Fig 8. It can be seen that the concentration in the raw water has been quite variable, ranging from 20 to 110 µg/l with an average of 93 µg/l. The concentration of TCE in the aerated water was below the WHO Guideline level of 30 µg/l on all occasions. The peak, at day 165, is likely to be an erroneous result: it is extremely difficult to maintain an adequate level of sample cleanliness when dealing with such low concentrations. The percentage removal graph, Fig 8 shows three such low points where the validity of the results must be questioned.

Aside from these three points, the removal efficiency ranges from 87-98%. The average removal efficiency is 92.9%, or 93.8% omitting the three low points.

If these removal rates are used in Equation (3) K_La can be shown to be 57.7.

When mass transfer is liquid rate controlled K_La can be calculated from the following equation (Norman W S, 1961).

$$K_La/D = 120 (L/\mu)^{0.75} (\mu/\rho D)0.5 \text{ (Imperial Units)} \qquad (4)$$

The diffusivity of TCE = $2.67.10^{-5}$ ft²/h . K_La is therefore calculated to be 72.8. It can be seen therefore that the actual mass transfer rates achieved are 80% of theoretical values. This is typical in that about 0.5 m of tower packing is usually considered as ineffective since it is functioning primarily as a water distribution section (at the top) and an air distributor at the bottom.
The removal efficiency for PCE averaged 93.6% when three apparently anomalous results were ignored.

3.5   Case Study B

In Case B two packed towers were constructed to treat 500 m³/h of groundwater contaminated with

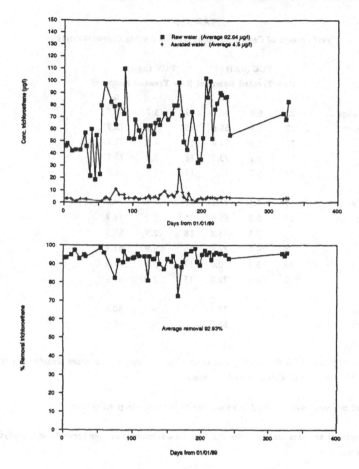

Fig 8 Removal of trichloroethene by Air Stripping (Case A)

TCE, PCE and low levels of CTC. The packing supplied by the Contractor was 94 mm Snowflake Type II and this was the first occasion that this has been used for such applications in the UK although experience exists in the USA. This packing has the benefit of a low pressure drop thereby allowing the full flow to pass through a single tower in the event of a fan failure.

As stated in Section 2 the well has a history of organic contamination and therefore activated carbon was installed after the stripping columns as a "polishing" stage.

Removal efficiencies specified were 88% for TCE 90%, PCE and 80% for CTC.

Commissioning

Poor removal efficiencies were observed during the commissioning of the works. As can be seen from the results in Table 2, the TCE removal was only 72% as compared with a design value of 88%.

Clearly the performance fell considerably short of the target with K_La from Eq 3 equal to 26.2.

The theoretical value of K_La expected for the towers as designed (calculated from Eq 4) was 99.5

TABLE 2

Performance of Case B Stripping Towers During Commissioning

		TCE (µg/l)			PCE (µg/l)		
		Raw	Treated	Removal %	Raw	Treated	Removal %
a)	One pump/	22	6.6	70.0	19	3.7	80.5
	one tower	22	6.9	68.6	18	3.8	78.9
		21	5.3	74.8	18	4.1	77.2
		21	5.1	75.7	18	3.7	79.4
		24	6.6	72.5	17	3.7	78.2
b)	Two pumps/	17	4.4	74.1	15	3.1	79.3
	two towers	17	5.2	69.4	15	3.2	78.7
		21	5.5	73.8	18	2.9	83.9
		18	4.9	72.8	16	2.9	81.9
		19	5.7	70.0	17	3.1	81.8
		18	4.9	72.8	15	2.8	81.3
	Average	-	-	72.2	-	-	80.1
	Design	-	-	88.0	-	-	90.0

which when compared with the actual value of 26.2 shows only a 26% mass transfer efficiency compared with the value of 80% achieved in Case A.

A number of measures were adopted in an attempt to improve the performance:

i)   Duplicate analyses confirmed that the problem was genuine and not created due to analytical errors.

ii)   In order to test whether poor distribution was the cause of the poor performance additional distributors were tried for air and water but no significant improvement in performance was achieved - the removal of TCE increased by only 1% to 73.9%.

Operation

Examination of the subsequent operating results in Fig 9 shows that the removal efficiency after commissioning rose to 90.64% when six low points were discarded.

From this data the operating K_La can be calculated to be 50.5, giving a mass transfer efficiency of 51% which shows a considerable improvement over the 26% achieved during commissioning.

It was thought that this performance improvement might have been due to the method of operation following commissioning. In order to follow demand, one well pump operates continuously whilst the second pump cuts in and out on demand but at all times the two towers are in service. The effect is that both pumps operate typically for only 10 hours per day. The time of sampling was not selected to coincide with periods when two pumps were operating.

The results have therefore been examined for three cases:

i)   Assuming 1 pump operating $L_m = 2682$ kg mole/m²/h

ii)   Assuming 2 pumps operating L_a = 5362 kg mole/m²/h

iii)   Flow equivalent 2 pumps 10 hours + 1 pump 14 hours L_a = 3799 kg/mole/m²/h

	Actual $K_L a$	Theoretical $K_L a$	Transfer Efficiency %
Case 1	23.5	59.1	40
Case 2	50.5	99.5	51
Case 3	34.3	77.2	44

All of these transfer efficiencies are considerably higher than that achieved during commissioning and it is therefore clear that a significant change in performance occurred between commissioning and operation.

It has long been known that etching of packing has the effect of increasing the specific surface area and therefore increasing mass transfer rates. Improvements in mass transfer performance of up to 50% have been observed from such effects (Norman and Soloman, 1959). Precipitation of calcium

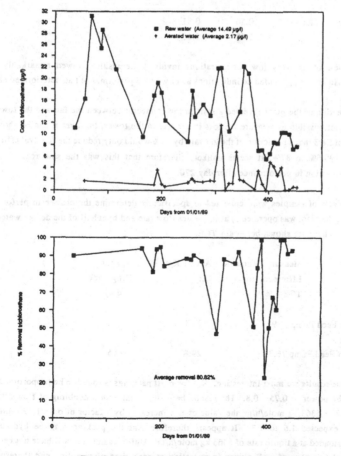

Fig 9 Removal of trichloroethene by Air Stripping (Case B)

carbonate scale upon packing surfaces might cause such improvements but the dosing of Calgon should prevent such deposition and therefore the formation of such roughened surfaces is unlikely.

No specific cause for the performance improvement was identified.

It can be seen from Fig 9 that the performance deteriorated significantly after day 270, falling to about 50% removal on a number of occasions (a single analysis showing 22% removal is not considered to be representative). Despite this deterioration in performance the treated water still met the target quality but nevertheless the deterioration in performance required investigation.

A number of factors have been studied in an attempt to explain the variation in performance, including:

i)   Duplicate samples were collected and analysed at the end of the period shown in Fig 9 in order to ensure that the problem was not due to analytical errors.

        Results for TCE were as follows (in µg/l):

        Raw water   8.53   :   7.58
                           7.76   :   7.58

        Aerated water   0.56   :   0.81
                           1.48   :   2.15

        In view of the very low concentrations involved the results showed remarkably good consistency and provided no indications as to why the performance had deteriorated earlier.

ii)   Inspection of the plant revealed a split in the connector between the fan and the tower. It was not possible to measure the loss in air rate to the tower; however it can be shown by use of Eq 3 that a reduction of the gas rate by 50% would only reduce the removal efficiency from 90.6% to 87%. It seems unlikely therefore that this was the sole reason for the deterioration in performance after day 270.

iii)   Two sets of samples were collected to specifically determine the change in performance when the plant was operated a) at full water flow rate and b) at half of the design water flow rate. These are shown below for TCE.

Measured Removal Efficiency for TCE - %	Actual K_La (Eq 3)	K_La for Ring Pack (Eq 4)
One Feed Pump 92	25.1	59.5
Two Feed Pump 76.3	29.8	99.5

These results are most interesting. K_La for most packings is found to be proportional to L_M to the power of 0.75 - 0.8. The results here show that with a doubling of flow rate from 2682 to 5363 kg mole/h/m² the value of K_La increases by a factor of only 1.19 rather than the expected 1.6 to 1.7. It appears therefore that the packing may be hydraulically overloaded at a liquid rate of 5363 kg mole/h/m². Unfortunately records have not been kept for the number of well pumps in operation at each time of sampling and therefore this explanation for the variation of performance throughout the operating period cannot be checked against the historic data.

4.   CONCLUSIONS

The case studies described in this paper have shown the need for thorough studies of the groundwater pollution problem in connection with the design and operation of treatment facilities and the need for a conservative approach to the treatment works design since there are still factors which are not fully understood.

Specific conclusions on the source identification aspect include:

- the importance of undertaking detailed solvent use surveys prior to siting exploratory boreholes;

- the difficulty in determining the age of pollution due to the presence of low permeability layers;

- the need for regular monitoring of a wide range of possible contaminants in industrial areas.

A number of conclusions have been drawn also from the experience in the use of air stripping for VOC removal at the two sites and possible reasons for the variability in performance are presented:

- air stripping has been effective at both sites in reducing the concentrations of TCE and PCE to target levels (12 µg/l and 5 µg/l respectively).

- at the Case A site where polypropylene pall rings were used as the tower packing material, performance was much as expected with 93% removal achieved for both TCE and PCE.

- at Case B site the performance has been more variable, for example with the removal efficiency for TCE varying between 50 and 92%.

- whilst this variability is not yet fully explained, the evaluation of operating data points to aspects which warrant further investigation, including, for example, the susceptibility of the Snowflake packing to hydraulic overloading with adverse effects on mass transfer performance.

- whilst the performance of activated carbon at the Case B site has not been covered in this paper, the installation of this second stage treatment following a history of trace organics in the raw water, has provided an additional safeguard against the breakthrough of VOCs.

ACKNOWLEDGEMENTS

The authors wish to thank the Director of Quality of Anglian Water Services Limited for permission to publish this paper. However the views expressed are those of the authors and are not necessarily those of Anglian Water Services Limited.

REFERENCES

Department of the Environment 1989 "The Water Supply (Water Quality) Regulations 1989". Statutory Instruments 1989 Nr 1147.

European Economic Community. 1976 "Directive Relating to Dangerous Substances in Water 76/464/EEC", Official Journal of the European Communities L129.

European Economic Community. 1980 "Directive Relating to the Quality of Water Intended for Human Consumption 80/778/EEC". Official Journal of the European Communities L229.

Narasimhan, T N & Witherspoon, PA, 1976. "An Integrated Finite Difference Method for Analysing Fluid Flow in Porous Media". Water Resources Research, 12(1), 57-64.

Green D W & Moloney J O, 1984. Perry's Chemical Engineers Handbook, 6th Edition, McGraw Hill.

Norman W S. 1961, Absorption, Distillation and Cooling Towers. Longmans.

Norman W S & Soloman B K. 1959. "The Effect of Ammonia Absorption on the Wetted Area of a Packed Tower", Trans Inst Chem Engrs 37, 237

Tester D J & Harker, R J 1981. "Groundwater Pollution Investigations in the Great Ouse Region". Water Pollution Control, 614-631.

US Environmental Protection Agency, 1975. "Preliminary Assessment of Suspected Carcinogens in Drinking Water", Report to Congress.

World Health Organisation, 1984. "Guidelines for Drinking Water Quality". WHO, Geneva.

PART VI

Rivers and river management

Chapter 21

GREEN ENGINEERING: THE CASE OF LAND DRAINAGE

J Pursglove (Mott MacDonald Ltd, Cambridge, UK)

SUMMARY

Engineers working on roads, rivers, dams, tunnels, bridges, landfill sites, quarries and buildings all have a profound effect on the environment. Some industrial landscapes, largely created by engineers in the canal age, for example, are now valued as exciting places to visit and enjoy. In our time, however, engineers have been rather more associated with the destruction of the environment. Roads have been and continue to be the subject of fierce environmental controversy. Rivers have been straightened out and their banks have been stripped of trees and concreted, as part of land drainage schemes. Waste management engineers have often assumed that landscapes which do not make a profit for their owners, such as woodlands and wetlands area automatically 'derelict' and therefore good places for dumping rubbish in.

These problems often arise because of the narrow brief which engineers are given. With a wide brief which includes the environment, they can even improve the environment. Land drainage schemes present an opportunity to plant trees and create ponds which were previously the victims of drainage. Roads, once care is taken to chose the best possible route for them, can create the opportunity to re-establish hedgerows and woodlands in the adjacent countryside through which they pass. The budget for this will be a tiny fraction of the engineering cost of the road building. Landfill sites may also create sufficient profit for a proportion to be ploughed back for the creation of adjacent habitat.

This 'green brief' for engineers is worthwhile because it contributes to the quality of life - something we all need as well as water for the reservoirs or an effective transport system resulting from road building. However, there are also plenty of hard-headed reasons for green engineering. Minimising on topsoil on road embankments saves money and creates ideal conditions for wild flowers. Trees planted on river banks as part of land drainage schemes both shade out aquatic weeds which would otherwise block the channel and help to reinforce eroding banks. Structural engineers involved in the construction of industrial estates are quick to grasp that a well landscaped 'science park' can command higher rates than a run-down collection of tin sheds. In addition there is always the real benefit of good public relations for a genuinely 'green' engineering scheme.

Of course there are occasions when the greenest engineering is really no engineering at all. This was the case in the proposals for dam building on the River Loire at Serre de la Fare, where the proposed reservoir would have destroyed a unique valley and no 'green design' could have ameliorated the

Water Treatment – Proceedings of the 1st International Conference, pp. 219–222

environmental disaster. However, I do believe that there are other areas on the Loire where the environment would be compatible with a carefully designed reservoir which is also sorely needed to reduce the real dangers of flooding.

The might of engineering machinery and the scale of engineering schemes is so considerable, that, if thoughtfully used, it can be one of the greatest powers for good in the protection and enhancement of the environment.

Engineers will be familiar with the arguments over the straightening and cementing of rivers to reduce flooding and the drainage of wetlands for agricultural production. In 1983, internal drainage boards were likening conservationists to 'Argentinians' while the boards were described equally unflatteringly as 'infamous'. Controversies over wetland drainage have sent ministers of state scurrying to some of our remotest and most inhospitable marshes. It is only just over five years since Mrs Thatcher herself became involved in the controversy over the Halvergate Marshes, while the drainage lobby were burning effigies of the conservationists on the Somerset Levels.

So what happened to the wicked drainage engineer? Has he really become an extinct species, as predicted by one Somerset farmer, observing the rise of the conservationists in 1986: 'And what's it done to the farmers and the drainage men? I'll tell you: we're the endangered species now. You should see us. We've all got copies of the Wildlife and Countryside Act 1981 at the bedside now, seeing if there is any mention in there of the Greater Crested Farmer or the Lesser Spotted Drainage Engineer. You look in any of these farmhouses around here at 11 at night and they'll all be reading a couple of chapters of the barmy thing before dropping off to sleep.'

Not entirely. The water authorities still carry out large drainage programmes, while county and district councils are busy on the 'non-main river' tributaries. If you cast a shrewd eye around the countryside in March you cannot fail to notice that spring is heralded not only by splashes of yellow celandines and pale primroses in the hedge banks, but also by the gleaming yellow diggers, hastily working their way up the watercourses to use up their allocated budget before the financial year ends on 1 April.

Sometimes the environmental results are as bad as ever. In 1988 the mature trees along the Fareham Brook south of Nottingham were felled and the stream gutted, in a scheme as crudely destructive as any I have ever seen. The fact that the Nature Conservancy Council - apparently misreading or not understanding the drawings presented to it by the internal drainage board - accepted the scheme, did not make the total destruction of riverside habitat any less catastrophic for the local wildlife.

RECENT LEGISLATIVE AND INSTITUTIONAL CHANGE

While the whole administrative structure of land drainage is being overhauled as part of privatisation, the opportunity has been missed to reform the notoriously anachronistic internal drainage boards, where voting strength at elections is proportional to the amount of property which an individual owns. At the other end of the drainage spectrum, Welsh farmers in Powys are doing their own thing on the River Severn, dredging up salmon spawning beds in a vain attempt to train the river not to erode their land. Yet such is the law, that it is hard for the water authority to prevent them from damaging the environment, unless they are also worsening the flooding.

By 1989, however, the general situation has enormously improved. Effective statutory protection for the Somerset Levels and other major wetlands has come in since 1984, a year which also saw a government inspired redirection of drainage away from rural agricultural schemes towards coastal and urban flood defence. In February 1988, EC leaders abolished the previously guaranteed price for wheat, signalling the end of the great corn boom, in a move of profound importance for the landscape. This was followed in July 1988 with Environmental Impact Assessment for land drainage schemes which have brought drainage under the first tentative stages of planning.

The setting up of the National Rivers Authority as part of the deal for water privatisation has to be good news for rivers and wetland landscapes, although the real power of the environmentalists as opposed to the drainage lobby within the NRA has yet to be tested.

CHANGING ATTITUDES AND METHODS

One major reason for what has amounted to a revolution in river management has been the many initiatives made by river engineers to take on a brief for the river landscape. The sheer undeniable beauty of river landscapes has been their own best advocate in bringing about a common realisation that a new kind of bridge building in river management is needed, involving engineers, digger-drivers, farmers and conservationists, to ensure that, which the risks of flooding are reduced, the wildlife and the wild quality of the rivers are preserved and even enhanced.

With this realisation came another discovery; that, subject to some critical qualifications concerning cases in which any drainage would be environmentally inappropriate, the whole, hitherto destructive, process of river engineering could now be turned on its head. By working to a brief which takes full account of the life-enhancing potential of a river, rather than its narrow definition as a drain, the erstwhile destroyer can now become a creator. It was recognised that machines which previously straightened every bend, filled every pond, and felled every tree could, at very little extra expense and at no sacrifice to drainage efficiency, actually extend riverain habitats for dragonflies, kingfishers, and otters, not to mention adding to the sum of human happiness for children with jam jars, walkers on riverside footpaths, and landowners out for an evening canter or in pursuit of duck and pheasant.

Mighty machines can now be seen transporting water forget-me-not to newly created margins at the water's edge, the emerald-green cushions starred with their delicate baby-blue flowers trembling over the edge of the great steel scoops before being laid gently down into a new niche of carefully prepared river ooze.

TECHNIQUES FOR CREATING NEW RIVER LANDSCAPES

The impact of machinery originally designed to destroy habitat can, when properly used, be nothing short of magical. A 'Swamp Dozer,' a machine so colossal that it literally makes the ground tremble, is able in a matter of days of pull back a whole field of builders' rubble or rye-grass to recreate the kind of swamp it was originally designed to destroy. In a few hours, it can hollow out a sizeable pond.

The response of the natural world to such activities may be immediate. In 1984 one of a dozen such ponds was created beside the Warwickshire Leam as part of a land-drainage scheme. Before the machine had actually finished, a not-so-shy moorhen was building its nest on the island which the driver had created for it as a relatively safe refuge from such nocturnal marauders as hedgehogs, which snuffle out moorhen eggs as a special delicacy. On a river near Evesham, the machines, as part of a flood alleviation scheme, extended a large bay and made a pool within the river bed. No sooner was the new pool created than the first swallows were dipping and swooping over it; and, best of all, by late afternoon of the first day, the bank top was crowded with bicycles thrown down in the grass by the village children while they enjoyed their new bathing-pool.

This kind of work has proved more interesting for the machine drivers who previously took a pride in the neatness of their straight channels, but who are now rising to the challenge of doing an 'untidy' job. Farmers, whose co-operation is essential in donating riverside land for tree planting, ponds and other habitat creation, have been universally generous, both on intensive arable land in the east and in the hill country in the west.

River engineers who take greater care and thought about the environment often find there are practical benefits which help the actual efficiency of the scheme. Ponds can sometimes be created in order to provide the material for raised flood banks, thereby saving the cost of importing spoil. Raised banks themselves can be spread with minimal top soil (another saving), and sown with low-growing grasses. This can reduce massive mowing costs, such as that currently incurred by the National Rivers Authority Severn Trent Region on 200 miles of flood bank alongside the River Trent. The environmental bonus will be many more wild flowers and butterflies.

Riverside trees will shade out aquatic weed and are among the best bank stabilisers. On the River Clwyd in North Wales, a river engineer came up with a particularly neat solution to a problem of bank erosion in 1977. A willow had fallen into the stream, and as a result, the river, nudged out of its regime, was eroding the adjacent bank. The engineer removed the offending tree, and chopped it into logs with which he filled mesh baskets, which he then set into the newly eroding bank. These grew immediately, efficiently holding the bank against erosion. Returning 10 years later, the engineer was able to admire a riverside grove of 20ft willows still attractively solving the problem which their parent tree had created.

CONCLUSION

Drainage schemes have been massively boosted by public money. For this reason it seems only fair that such schemes respect and even enhance the common heritage of the landscape, and since 1981 this has indeed become a legal responsibility of drainage engineers. Currently, the National Rivers Authority Severn Trent Region spends a mere 5% of its total annual river maintenance budget on just such endeavours.

If everyone responsible for the maintenance of watercourses rises to such a proportionally modest commitment, then the loveliest of our existing rivers will not be lost, and all the miles of denuded drain which lie waiting, like sleeping beauties in intensive care, will, over the space of a generation, recover life and elegance as they carry away their essential cargo of flood-water to the open sea.

Chapter 22

PREVENTION OF WATER POLLUTION ARISING FROM NON-POINT DISCHARGES FROM URBAN AREAS

V M Khrat (All-Union Scientific Research Institute for Water Protection, Kharkov, USSR)

SUMMARY

Detrimental effect of nonpoint (NP) discharge from urban areas on Biolochemical Oxygen Demand (BOD) and Dissolved Oxygen (DO) of streams and water bodies was evaluated by a mathematical model on the basis of field data and computerized prognostic calculations. Various situations were analysed at different combinations of geomorphologic characteristics of water bodies and different dimensions of populated areas. As a result there were defined water bodies needing urgent protection against pollution from urban NP discharge. Analysis of economic indices of traditional methods of urban stormwater treatment, on special treatment units with stormwater discharge systems, demonstrated high cost of such measures, especially if expressed in units of mass of removed pollutants. This led to a conclusion that it is impossible to solve the problem in the near future along these lines only.

A so-called complex ecological and technological approach to solving the problem is offered. It involves quality control in the process of the NP discharge formation by introduction of normative limits on pollutants building up in urban areas. The limits should be observed by limiting industrial emissions to the atmosphere, soil erosion, destruction of road coverings and by the improvement of ecological situation as a whole. On territories with heavy technogenic loadings (industrial zones, highways, etc) it is necessary to construct local stormwater discharge and treatment units for first portions of NP discharge.

Such measures will not only limit pollutant discharges (currently tending to increase as a result of urbanization) but also allow cuts of 25 - 30 % compared with the present level.

© 1991 Elsevier Science Publishers Ltd, England
Water Treatment – Proceedings of the 1st International Conference, pp. 223–230

INTRODUCTION

The present practice of protection against water pollution arising from urban NP discharges concentrates mainly on construction of expensive discharge systems providing for entrapment of stormwaters and their treatment as a combined effluent, or on special treatment units at the stormwater sewer outfalls. However this would demand - for the country as a whole - capital investments which, according to scoping calculations, are comparable with the present value of the main assets of water protection (khvat, 1986). Moreover, at present there are no scientifically substantiated data for evaluation of the final efficiency of the investments as to improvement of the natural waters. So it is obvious that the traditional approach to the solution of the problem must be re-examined.

A new concept of protection of water resouces against pollution from NP discharge must be based on objective data, defining a degree of the negative effect of this pollution source on quality of natural waters. It is necessary to provide for a stage-by-stage solution of the problem, with application of economically viable water protective measures.

Methods

At present the mechanism of urban NP discharge action on water bodies is not sufficiently studied. However a range of negative factors, such as Suspended Solids (SS) and BOD increase in the river water and DO decrease during rainstorms are fairly obvious and amenable to quantitative evaluation.

It must be stressed that the worst situation in the stream may occur not at points of direct stormwater discharge in the city limits but below the point of treated municipal effluents discharge where these two powerful sources of pollution intensify each other (Nevzorov, 1986).

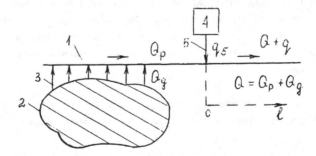

Fig. 1 A diagram of urban NP discharge inflow to the stream.
1 - stream, 2 - city, 3 - stormwater outlets, 4 - sewage treatment works, 5 - treated effluents outlet

A design diagram on which the mathematical model is based is presented in Fig 1. It permits evaluation of the influence of NP discharge on BOD and DO in the river water at the most polluted river segment below the treated effluents outlet.

DO concentration along the river was determined by the following equations:

$$S(\ell) = a_r(a - S_\varphi) \cdot 10^{-K_p \frac{\ell}{V_i}} - K_{CM} \left[\frac{\mathcal{L}_{cr} \, g_\varepsilon + \delta' Q \mathcal{L}_\varphi}{(g + \delta' Q)(K_p - K_{CM})} \right] \times$$

$$\times \left(10^{-K_{CM} \frac{\ell}{V_i}} - 10^{-K_p \frac{\ell}{V_i}} \right) + \frac{g_\varepsilon \cdot S_{cr}}{g_\varepsilon + \delta' Q}, \tag{1}$$

$$K_{CM} = \frac{V_i}{\ell} \left[\frac{\mathcal{L}_{cr} + (n-1) \mathcal{L}_\varphi}{\mathcal{L}_{cr} \cdot 10^{-K_{cr} \frac{\ell}{V_i}} + (n-1) \mathcal{L}_\varphi \cdot 10^{-K_\varphi \frac{\ell}{V_i}}} \right], \tag{2}$$

where $S(\ell)$ - DO concentration at a distance $\ell(m)$ from the sewage treatment works, mg O_2/l; a - limiting solubility of oxygen in water at a given temperature, mg O_2/l; K_p - reaeration coefficient, days^{-1}; K_{cr}, K_φ - constants of oxygen consumption by effluents and river water organics, days^{-1}; S_{cr}, S_φ - oxygen concentration in effluents and the river O_2 background above the sewage treatment works, mg O_2/l.

$$S_\varphi = \frac{S_g Q_g + S_p Q_p}{Q_g + Q_p}, \tag{3}$$

S_g, S_p - oxygen concentration in NP discharge and in the river above city, mg O_2/l; Q_g, Q_p rates of discharge of NP and the river above the NP inflow m^3/s; Q - sum of the river and NP discharges $Q_p + Q_g$, m^3/s; $\mathcal{L}_{cr}$, $\mathcal{L}_\varphi$ - BOD$_{ult}$ of effluents and the river water and NP mixture above the sewage treatment works, mg O_2/l;

$$\mathcal{L}_\varphi = \frac{\mathcal{L}_g Q_g + \mathcal{L}_p Q_p}{Q_g + Q_p}, \tag{4}$$

$\mathcal{L}_g$, $\mathcal{L}_p$ -BOD$_{ult}$ of the NP discharge and the river water prior to NP inflow, mg O_2/l; g_ε- effluent rate, m^3/s; V and V_i - rate of river flow below the sewage treatment works, in m/s and m/day, respectively:

$$K_\varphi = \frac{K_p Q_p + K_g Q_g}{Q_p + Q_g}, \tag{5}$$

K_g - constant of NP oxygen consumption rate, day^{-1}; n - effluents dilution ratio, δ' - coefficient of mixing.

Experimental data allowed to assign $K_g = 0.3$ day^{-1}. Computing produced a file applicable to evaluation of BOD and DO in the river water in a wide range of Q_p, Q_g, g_ε, V, K_p, a, S_φ, $\mathcal{L}_\varphi$, $\mathcal{L}_g$ combinations.

The rate and BOD of NP discharge were set corresponding to towns with populations ranging from 50000 to one million inhabitants. The design storm was set at 12 mm rain, providing for 90% wash off of solid particles built-up on impervious surfaces accessible to the rain flow (Khvat, 1988).

Specific wash-off of organics by the design storm per one inhabitant is a function of the town population (Khvat, 1988).

A rectangular averaged hydrograph analysis for the whole town area was performed on the basis of predetermined relation of the types of territory areas to the number of inhabitants, with allowance for a given duration of rain, rate of flow and, corresponding travel time of the flow to the control point. The rate of municipal effluents was set in accordance with normative volumes per one inhabitant.

Results

The inflow of NP discharge into the river gives a considerable impact on the BOD of the river water. In rivers with discharge from one to 50 m^3/s during rain storms the river water BOD can increase 1.1 to 16 times, depending upon the town dimensions. However at present it is difficult to evaluate detrimental effects of such temporary overstepping of the BOD norms. At the same time, if BOD increase is concurrent with DO decrease in the stream, the situation must be considered as extremely dangerous. Under such conditions it is necessary to provide for urgent implementation of measures of NP discharge treatment. Calculations indicate that rivers with natural hydrological and morphological regime subjected to a temporary BOD increase caused by rain storms, as a rule maintain normal DO regime. The situation would be much worse in rivers with low flowage, up to 25 m^3/s and low flow rates, below 0.3 m/s.

On the diagram (Fig 2) the area I indicates the most unfavourable combination of rivers; flow rates and dimensions of towns situated in their basins. In the area with the river flow rate 1 - 25 m^3/s and many towns with population from 50000 to one million inhabitants, 3 - 5-fold increase of the river water BOD as compared to the normative value, DO regime deterioration is observed.

Presently in the majority of towns streams are regulated and they are practically systems of reservoirs with very low flowage and rates of flow.

A formula is offered for rating DO regime under such conditions on the basis of the oxygen balance equation, with allowance for the bottom sediments:

$$C_t = a - (a - b) \cdot 10^{-K_2 t} - \frac{K_1 \, L_a}{K_2 - K_1} \left(10^{-K_1 t} - 10^{-K_2 t} \right) -$$
$$- \frac{A F_g \cdot 0,434}{V K_2} \left(1 - 10^{-K_2 t} \right), \tag{6}$$

where C_t - oxygen concentration in the mixture of NP discharge and reservoir water afret time t, mg O_2/l; b, L_a, K_1 - initital values of DO in the mixture, BOD mg O_2/l, and

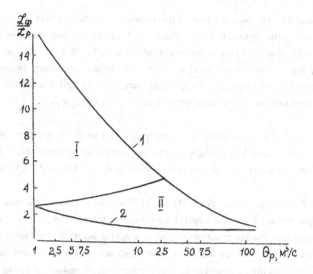

Fig 2   Influence of urban NP discharge from different numbers of inhabitants on BOD and DO in the stream.

relationship of river and NP water mixture BOD to the river water BOD;   - stream discharge, m^3/s, I and II - steam segments with reduced and normal DO, correspondingly.

organics oxidation rate constant day^{-1}, respectively; A - bottom sediments oxygen consumption rate mg/m^2 per day; f_g - bottom sediments area, m^2; V - volume of NP and reservoir water mixutre, m^3.

The minimal DO concentration will occur after time determined by formula

$$t_{\kappa\rho} = \frac{\lg\left\{\frac{K_2}{K_1}\left[1-\frac{(a-b)(K_2-K_1)}{K_1 L_a}+\frac{A \cdot f_g}{V}\cdot\frac{(K_2-K_1)\cdot 0,434}{K_1 K_2 L_a}\right]\right\}}{K_2-K_1} \qquad (7$$

Substituting (7) into (6), C_{min} can be found and compared with the normative DO value.

Field investigations conducted in the city of Kharkov elucidated features of oxygen consumption by the solid phase in the discharge (Promá, 1983) and by bottom sediments originating primarily from NP discharge (Khvat, 1983).

Computer calculations, performed on the basis of (6) and (7), evaluated the influence of NP discharge on DO regime in reservoirs and sluggish streams in the wide range of volumes of NP and river water in the town segments of the water body, as well as an initial oxygen deficit and average depths of the reservoir. Field data of the bottom sediments oxidation rates from 0.34 to 4 g/m^2 per day were used in calculations.

The analysis applied to streams and reservoirs allowed definition of water bodies needing urgent protection against pollution from urban NP discharges. Such are streams of low water content, with minimal rates in the warm seasons up to 30 m^3/s, low rates of flow, below 0.3 m/s, and specific river discharge per one urban inhabitant below 0.03 l/s. Urban reservoirs with low flowage, with specific water volume less than 1000m^3 per 1 ha of built-up adjoined territories, can also be included in this category.

Traditionally SS loads can be reduced by construction of a partially separate system with partial bypass to the municipal sewage treatment works, or sedimentation lagoons at swere outfalls, and in exceptional cases filters can be provided for tertiary treatment.

Treatment of NP discharge gives relatively small BOD reduction, only about 3 - 6 % of the raw municipal sewage BOD, but capital investments for this purpose are comparable with that for municipal sewage treatment works. Capital investments even for simplest NP discharge treatment, such as lagoons, when calculated per one tone of captured organics, are 2 - 3 times higher than for tertiary treatment of municipal sewage.

Such high specific costs, especially capital ones, for NP discharge treatment units are primarily due to the NP irregular periodicity.

At present the main sources of urban NP discharge pollution are sedimenting aerosols, destruction of road coverings, tyres attrition, and soil erosion. If erosion products were reduced to average values, then polluting loads of the aerosol component would amount to 58% for cities and to 78% for towns. Products of road covering destruction amount to 10 - 25 %.

Investigation of dispersive and dynamic characteristics of urban aerosols obtained data on SS concentration in NP discharge produced by the design rain under various degrees of dust content in the air (see Table).

The table data demonstrate that reduction of dust content in the lower atmosphere, primarily in areas with heavy polluting industries, can alone cut SS wash-off by several times. Such an approach is economically viable.

According to our calculations capital costs of 1kg of dust removal from the air is about 85-fold cheaper than 1 kg of SS removal at a partially separate system unit for NP discharge treatment and about 16-fold cheaper than removal in lagoons.

Prognostication demonstrated that at the normative air dust content 0.15 mg/m^3, prevention by simple measures of soil wash-off from lawns and open areas, and higher durability of road covers can reduce by 30% the total wash-off of pollutants caused by the design rain.

Application of environment protecting and technical measures reducing technogenic polluting loads in urban NP discharge will allow cuts in treatment costs of 33 - 70 %, and in some cases, with a large discharge of the stream, will completely obviate them.

Table - Sedimenting aerosols wash-off by NP discharge as a function of dust content in the air.

A	B	C	D	E	F
Clear slightly polluted	farms, non-industrial territories	0.15	0.8	1.8	0.2
	dwelling regions in industrial towns	0.5	2.6	6.0	0.6
Heavy polluted	urban industrial regions	1.0	5.0	11.3	1.3
Superheavy polluted	territories of heavy polluting industries	3.0	15.0	34.0	4.0

Designations: A - degree of atmospheric air pollution; B - area characteristics; C - daily average air dust concentration, mg/m^3; D - dust compiling intensity, g/m^2 per day; E - wash off by design rain, g/m^2; F - sedimenting aerosols concentration in NP discharge, g/l.

Conclusions

Introduction of limits on build up of fine-grained particles on urban territories is a promising way for environmental improvement as a whole in all media, including water bodies. If existing environmental protection measures do not reduce total load down to the limiting level, a programme of NP treatment units must be considered. Design of such units can be based on the balance equation of fine particles build-up and on found relationships (Khvat, 1988), allowing quantitative evaluation of technogenic components (atmosphere pollution, road cover destruction, soil erosion, etc).

Where urban territory is very heterogenerous it is necessary to solve an optimization problem, to choose between alternative ways of reducing technogenic polluting loads (minimization of environment-protection costs, giving a prescribed quality of NP discharge in the water-use section).

References

Khvat V M (1983), Reglamentatziya vypuska poverkhnostnogo stoka stoka s gorodskoi territorii v vodnye objekty. Okhrana vod ot zagryazneniya poverkhnostnym stokom, Sc. proc. VNIIVO, Kharkov, P 3 - 12.

Khvat V M (1986), Planirovanie vodookhrannykh meropriyatii s uchetom poverkhnostnogo stoka, otvodimogo s zastroennykh territorii. Report abstracts of 2nd All-Union Conference: Sovershenstvovanie metodologii upravleniya sozialistichaskim prirodopolzovaniem, v, II, P 319 - 321.

Khvat V M (1988), Analiz antropogennogo vozdeistviya na formirovanie poverkhnostnogo stoka gorodov. Modelirovanie i kontrol kachestva vod. Sc. proc. VNIIVO, Kharkov, p 80 - 89.

Nevzorov M I, Nechaevsky M L (1986), Vliyanie neorganizovannogo postupleniya poverkhnostnogo stoka na effectivnost raboty kommunalnykh ochistnykh sooruzhenii. Report Abstracts of the All-Union Scientific and Technical Conference Osnovnye napravleniya razvitiya vodosnabzheniya, ochistki prirodnykh i stochnykh vod i obrabotka osadka, part II, Kharkov, p 448 - 450.

Priina N G, Goryainov E I, et al. (1983), Vliyanie osazhdayuscheisya tverdoi fazy poverkhnostnogo stoka s gorodskoi territorii na kislorodnyi rezhim malykh rek. Okhrana vod ot zagryazneyiya poverkhnostnym stokom. Sc. proc. VNIIVO, Kharkov, p 37 - 41.

Chapter 23

THE RIVER FANE FLOW REGULATION SCHEME: A CASE STUDY IN RIVER ENVIRONMENT MANAGEMENT

M F Garrick (Patrick J Tobin & Company Ltd, Galway, Rep. of Ireland)

1. INTRODUCTION

1.1 Design Concept

The River Fane Scheme is a Low Flow Augmentation Scheme which depends on the development of Lough Muckno as a Storage Reservoir in order to supply the water demand of Dundalk, the Republic of Ireland's sixth largest urban centre located in the North Eastern part of the Country. Water from Lough Muckno will supplement flow deficiency in the River Fane so as to provide the required flows at various points downstream and particularly at the Dundalk U.D.C. Intake at Stephenstown County Louth. The Scheme therefore has two centres of activity, the Headworks at Lough Muckno and the Intake-Treatment Works at Stephenstown and Cavan Hill, more than 20 km downstream.

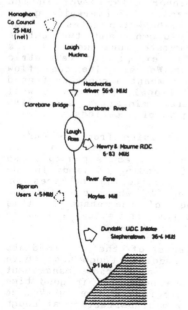

FIG 1   SCHEMATIC DIAGRAM OF HEADWORKS AND ABSTRACTIONS

Water Treatment – Proceedings of the 1st International Conference, pp. 231–242

The Scheme has been designed not only to meet the projected demand of Dundalk and its hinterland, but also to provide for projected requirements of a second Local Authority (County Monaghan) abstracting directly from Lough Muckno; of Newry and Mourne District Council in Northern Ireland who already abstract from Lough Ross (3km downstream) as well as Riparian Users and Compensation Water below the Stephenstown Intake.

The compensation water is intended to facilitate movement of migratory salmon, sea-trout and eels along the Fane River System. The Fane though relatively small in catchment area (350 sq.km) is one of Irelands primary late run salmon rivers, with adult fish moving upstream in September-November to spawning grounds in the Clarebane River, a short reach between Loughs Muckno and Ross, as well as to tributary streams in the upper catchment.

While Dundalk U.D.C. have a Water Order for a 36.4 Ml/d. abstraction at Stephenstown, the total Regulation Flow downstream of Lough Muckno will include all downstream water demand and river management requirements and will ultimately reach 56.8 Ml/d. The natural unregulated river has gone dry in 1975 and has a record - averaged 90 Percentile Flow of 17.3 Ml/d at a Gauging Station in the lower half of the catchment.

1.2   The Influence of the
River Environment on Headworks Design

The Scheme is unusual in that, in order to avoid irreparable damage to salmon spawning grounds by excavation in the gravel redds in the Clarebane River, a decision was made to design a Headworks that would regulate outflow from Lough Muckno in two ways; by gravitational spillage for as long in the year as possible, but then by pumping from a reduced water level in the upstream Lough Muckno to a higher water level in the undisturbed Clarebane River downstream. The overspill gates in Spring will therefore function as backstop gates for the pumped flow in late Summer and Autumn, and the entire installation is designed to operate unmanned, by a Programmable Logic Controller (PLC), except for telemetric supervision from the Main Treatment Works. The dual flow aspect of the Head Station design has meant a consequent need for a conventional two stage gravitational fish pass as well as a pumped flow Fish Chute for upstream migration and a Fish Lift for downstream migration during pumping periods.

A secondary feature of the Scheme, arising from the under-reservoired nature of Lough Muckno, is the impoundment of storage at as late a stage in Springtime, as is possible, so as to reduce the risk of additional flooding of lands in the upper catchment. This is achieved by monitoring the recession hydrograph of winter flows using the PLC which operates the control gates, and using knowledge of the shape of this hydrograph to manage the storage so that it is just full when it is first required to supplement river flow.

Thirdly the flow attenuation effects of the intermediate catchment and particularly the intermediate Lough Ross, have been studied and incorporated into a storage management policy, which adjusts releases at the Headworks in good time to match receding or rising flow hydrograph at the Intake, so as to reduce Environmental Impact by overabstraction at Lough Muckno.

The Scheme is also notable in that Simulated Gate Control Policies (1) were applied to historically wet and dry years to assist in explaining the environmental impact of the Scheme to

the public at the largest Public Inquiry ever held under
Ireland's Water Supplies Act and these policies were
subsequently used to evolve simple, Fail-Safe control software
particular to the requirements of the Site.

Finally, and in keeping with the non intrusive philosophy of
the Scheme, the architectural treatment of the Pumping Station
has provided a structure similar to a late 19th century
granary typical of the Lough Muckno area and fully
complementary to the Lake/River Environment there.

2.   HYDROLOGY OF THE LOUGH MUCKNO – FANE RIVER SYSTEM

The design of the Headworks at Lough Muckno followed a
detailed Hydrographic Survey of the Lake bed as well as
hydrological study of the Lake-River System.

As has been mentioned, there are two lakes of significant size
on the Fane River System, Lough Muckno being the upstream and
major lake, connected with the downstream Lough Ross by a 3 km
short reach of river known as the Clarebane River. There are
two principal Flow Gauging Stations on the catchment, one at
Clarebane Bridge approximately midway between the two lakes,
the second on the River Fane downstream of Lough Ross at
Moyles Mill near Inniskeen, County Monaghan.

The principal characteristics of Lough Muckno and the flow
recording stations are as follows:

Lough Muckno

A.   STORAGE/AREA

1.	Lake area at Maximum Recorded Flood Level	568 ha
2.	Lake area at Lowest Recorded Summer Level	338 ha
3.	Impounded Storage (92.1m OD to 91.3m OD)	2.6 million cu.m
4.	Available Storage for Pumping (91.3mOD – 88.6m OD)	8.4 million cu.m

B.   LEVELS (metres AOD)

1.	Highest Recorded Flood Level	92.87
2.	Upper Impoundment Level	92.1
3.	Level at which Pumping must commence at full demand	91.3
4.	Lowest Recorded Summer Level	90.97
5.	Control Structure Apron Level	90.7
6.	Projected Drawdown Level at Full Demand in 1975 type Drought Conditons	88.6

TABLE 1 : **LOUGH MUCKNO HYDROGRAPHIC CHARACTERISTICS**

Flow Gauging Station	Clarebane Bridge	Moyles Mill
Catchment Area to Station	163 Km²	230 Km²
Mean Annual Flow	2.75 m³/s	3.76 m³/s
Peak Flood	17.1 m³/s	34.6 m³/s
Lowest Recorded Flow	0.02 m³/s	0.00 m³/s

TABLE 2 : **RIVER FANE FLOW GAUGING STATIONS**

Lough Ross is a relatively small lake, 88 ha in area, and will
operate without controls, rising and falling in response to
releases at Lough Muckno. Referring to the flow records, it
might appear anomalous that a flow of zero could be recorded

at a downstream station during a period when the flow at an
upstream station was 0.02 m³/s (1.7 Ml/d), approximately, but
the circumstances may be explained by the fact that a water
supply from Lough Ross is at present abstracted to serve the
adjacent area in Northern Ireland.

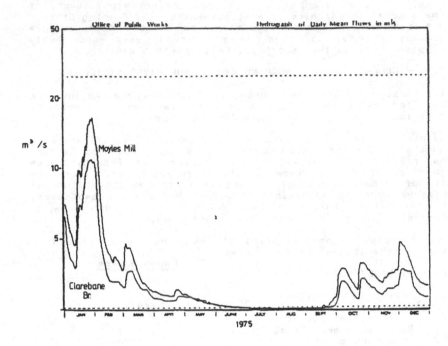

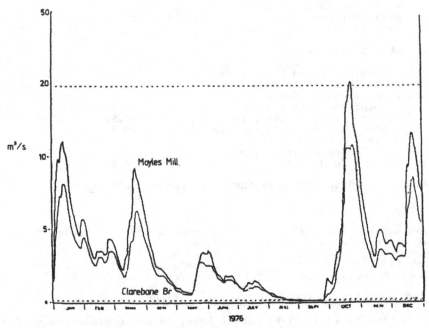

FIG. 2   HYDROGRAPHS AT CLAREBANE BRIDGE AND MOYLES MILL
          SHOWING LOW INTERMEDIATE BASEFLOW

It is a characteristic of the flows on the Fane System that although the Mean Annual Flow is respectably high, flows at or above this level do not persist over anything like half the year, and much lower flows are experienced over long periods in drier summers. The geology of the catchment is mostly Ordovician-Silurian Shales with some Sandstone Drifts and the Fane Catchment shares with the nearby Glyde and Dee Catchments a particularly low baseflow contribution from groundwater in prolonged dry weather (Fig 2).

Since the headworks are some distance upstream of the abstraction point, our initial assumption was that intermediate catchment flow could be counted upon to permit fractional releases at Lough Muckno. This assumption was tested by comparison of the flows at Clarebane Bridge and downstream at Moyles Mill over different periods of the year. When adjustment was made for the existing water supply abstraction from Lough Ross it was found that the overall correspondence between Catchment Area and Mean Annual Flow did not persist into the drier periods of the year. In 1975 for example 97% of the flow at Moyles Mill in May was already in the river at Clarebane Bridge and in the following month the flows at both stations were equal (when allowance for Newry is made). In June 1976 the figure was 87% and in May 1977 it was 91%. In designing releases from Lough Muckno for abstraction at Stephenstown during severe drought periods it was therefore assumed that the intermediate catchment would contribute negligible amounts to the overall flow arriving at the Intake.

3.   HEADWORKS DESIGN

3.1 Pumping Station Design

Based on a Hydrographic Survey of Lough Muckno, a Storage/Elevation curve for the Lake was prepared prior to the Water Order Enquiry of 1982. The flow records for 1975 & '76 suggested that in particularly dry years, with all Water Supply Authorities drawing at full demand, water level on Lough Muckno could drop to 88.6m above OD compared to normal summer levels of 91.0m OD. Since the Pumping Station is located approximately 450m from the lake outlet, and allowing for channel losses it became clear that a direct access for water from the 88m OD contour in the lake to the Pumping Station would be required and that the Wet Wells in the Station itself would have to anticipate a maximum drawdown water level of 88.0m OD at the Station. The construction of a lined and normally submerged concrete Revetment Channel in the bed of the Clarebane River between Lough Muckno and the Pumping Station therefore formed part of the Civil Works Contract.

The Pumping Rate depends on the flow defecit between water demand and the naturally available flow in the Fane at Stephenstown, so that the Pumps are variable speed pumps capable of pumping over a flow range from 11.4 Ml/d. to 56.8 Ml/d.

3.2 Gate Control

Activation of the Gates from the passive Winter position will take place on reaching a predetermined point on a falling recession hydrograph, or on reaching a predetermined date. The PLC will control the raising of the Gates so as to pass the Regulation Flow while impounding into storage, aiming to just achieve a full storage when drafts from storage are first required. Water Level upstream and downstream is logged at 15 minute intervals and the Gates pause at each movement (for a period which is related to the response time of the Clarebane River) during which the effect of the adjustment is monitored.

The PLC is programmed to drop the gates at a rate related to the known rate of rise of the worst recorded flood hydrograph on Lough Muckno, and when the spilling of excess floodwater has stabilised water level on the lake the programme will recognise when to resume impoundment of water. The Gate Control Panel has a backup battery power supply, as does the PLC and hydraulic accumulators are designed to permit one full manual cycle of adjustment of all the gates following power failure. The Gates will either remain in the vertical back stop position or will drop to the passive position whichever is fail-safe (at that time of the year) on power failure.

The hydraulic control of outflow from Lough Muckno will be balanced between free discharge over the gates on the one hand and backwatering of the Control Structure because of the throttling effect of the Clarebane River at higher flows on the other. Upstream water level determines the controlling head for free flow over the gates, downstream water level is then monitored together with the Rating Curve as a check on outflow. Once the flow over the gates is no longer free, control is passed to the downstream level alone.

4.   RECESSION CURVE ANALYSIS

Since Lough Ross lies between the headworks and the intake, and since these are in any case separated by more than 20km of the river length, it is clear that any variation in release from Lough Muckno, particularly at low flows, would be extensively attenuated in Lough Ross and in the river reach downstream. If water is not to be needlessly pumped at Lough Muckno, or if a deficit is not to arise at Stephenstown then releases at Lough Muckno will have to be anticipatory in nature both as to timing and magnitude. Long term prospects for river bank storage at Stephenstown will permit regulation of bank-side storage to compensate to some degree for over or under releases but a careful analysis of river recession rates will permit a substantial degree of control in the early years of the Scheme as demand grows towards its long term values.

Recession analysis depends upon the isolation of good recession hydrographs, uncontaminated by rainfall if possible, from which to derive the natural decay factor defined by the equation:

$$\frac{dQ}{dt} = -kQ \tag{1}$$

where k is a constant (the recession rate). The solution to this equation is given by

$$Q_t = Q_o e^{-kt} \tag{2}$$

so that the ratio of flows one time unit apart can be expressed as

$$Q_{t+1}/Q_t = R_Q, \text{ where } R_Q = e^{-k} \tag{3}$$

The assumption that the ratio of flows in successive time periods remains constant throughout the range of flow is often not justified in practice.

The difficulties of estimating the decay factor from the ratio of flows one day apart are many including errors of measurement, "contamination" of the recession by small amounts of rainfall, and in the case of Lough Muckno poor rating curve definition at very low flows. It was recognised moreover that the time required for equilibrium to be established between inflows to Lough Ross and outflows to the River Fane below would vary between 3 and 5 days approximately, with the longer period applying at low flows. Accordingly it was decided to estimate recession factor using the expression

$$\left(Q_{t+5}/Q_t\right)^{0.2} = R_Q \qquad (4)$$

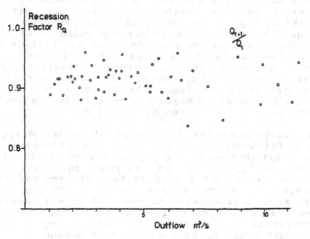

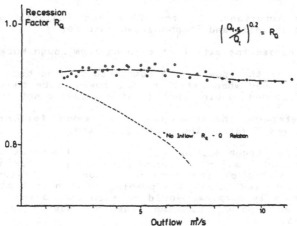

FIG 3   RECESSION FACTOR AT CLAREBANE BRIDGE

and these have been plotted for Clarebane Bridge in Fig 3. These show that the Recession Factor (R_Q) at Clarebane Bridge is relatively constant at 0.9 but rises to 0.92 at low flows. Since Storage/Yield analyses suggest that winter inflow will always be adequate to replenish storage at the levels of demand which are projected, inactivation of outlet controls for the winter period will be possible thus minimising interferance with winter floods.

A knowledge of the recession characteristics of Lough Muckno has permitted a point to be identified on the recession curve which will trigger the filling of storage to any selected target level while guaranteeing releases throughout the filling sequence. It remains to check that the recession factor has not been significantly altered by the outlet works which have been constructed over the past two years, since other sources (3) have found this to be the case elsewhere.

The presence of substantial natural storage in Lough Muckno greatly influences the recession rates on the hydrograph at gauging stations downstream. From the Storage/Elevation curve on Lough Muckno, and the rating curve for outflow from the lake into the Clarebane River, it is possible to calculate an approximate decay curve for the lake alone assuming no inflow from the minor lakes and catchment upstream, and this has also been plotted on Fig 3. In future it is the behaviour of Lough Ross at low flow that will dominate the response of the lower catchment during regulation periods since all inflows to Lough Muckno will be completely absorbed in replenishing storage depletion. While an accurate Storage/Elevation curve for Lough Ross has not been produced, there are indications that its "noinflow" recession factor is of the order of 0.66 for flows in the region of 20 Ml/d, and that it is somewhat larger (0.74) for flows in the region of 7.5 Ml/d.

Recession Analysis of the hydrograph at Moyles Mill, in the Lower Fane Catchment, suggests that the recession at this point is still dominated by Lough Muckno particularly at medium and high flows. We are now researching the extent to which the recession factor of 0.88 to 0.92 recorded at Moyles Mill changes as Lough Muckno becomes a more regulated reservoir. We have prudently assumed that it will approach the "no-inflow" calculated recession for Lough Ross but this can not be practically tested until the system is again exposed to a drought which requires pumping on Lough Muckno.

We attach importance to the Recession Analysis at Lough Muckno, Moyles Mill and Stephenstown in order

(a) to minimise the extent of pumping from Lough Muckno and

(b) to ensure that pumping is commenced in good time and then discontinued when refilling has taken place to an extent that resumed pumping shortly afterwards is not required.

(c) to determine the consequences of power failure on the Clarebane River during pumping periods.

The Station at Lough Muckno will be unmanned and the intention is to carry out a major maintenance cleaning of screens, pumps etc. at the end of the pumping season. The sequence of operation required to set the pumping regimen in train, while not particularly complex, would be such that a well defined, reliable end point to the pumping season should be established.

5.   MIGRATORY FISH MOVEMENT

The Control Structure includes two Fish Pass Gates which will in the initial stages of a drought simply act to divide the level difference between lake and river into two steps whilst flow to the Clarebane River spills over the gates. The PLC will control gate lip level so that a minimum depth of flow of 200mm over the Fish Pass Gates always exists, irrespective of minimum Regulation Flow.

Once pumping commences, a fraction of the pumped flow will be
delivered to a Fish Chamber which discharges over the
downstream Fish Pass Gate (Fig 5). Fish migrating upstream
will be attracted into the Fish Chamber and counted as they
enter. On reaching a pre-determined number of fish, or
overridden by a timer, a penstock will open allowing the fish
to be flushed down a chute which terminates upstream of the
Intake.

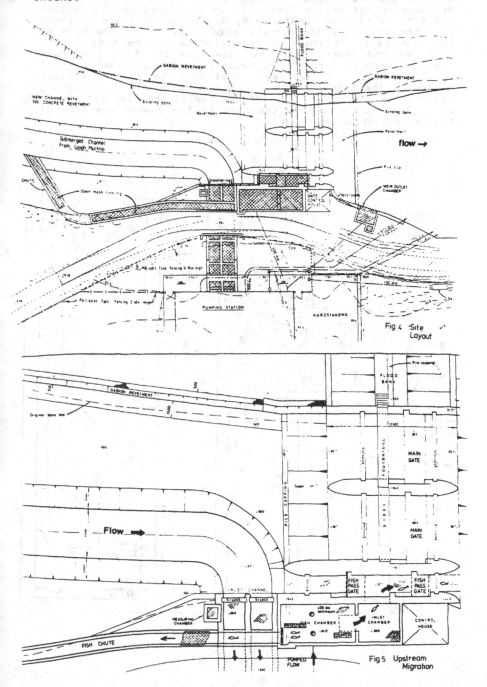

Fig 4 Site Layout

Fig 5 Upstream Migration

Fish migrating downstream during the pumping phase of operation must pass from a low level Lough Muckno to a higher level Clarebane River.   On entering the Station via the operating Intake Channel, they can pass through the coarse screens but are confined in the sump outside the Bandscreens. The Intake, Bandscreens and entry to the Wet Well are designed on a Twin Stream basis symmetrical about the axis X-X (Fig 6) so that any one half of the Intake-Bandscreen arrangement can accommodate the entire flow.   It is then possible to isolate one side by closing Penstocks A, B & C and pump its contents into the other, in the process drawing fish into the base of the Fish Lift.   Closure of the penstock at the base of the Lift (D) and reversal of the flow from the Fish Lift Pump together with closure of Sl. Valve E will flood the shaft of the Fish Lift.   A stainless steel mesh basket on vertical guides in the Fish Lift is then drawn up from below, forcing the fish to rise above it . At the top of the Fish Lift Shaft an outlet pipe will take the Fish to the downstream side of the Control Structure.

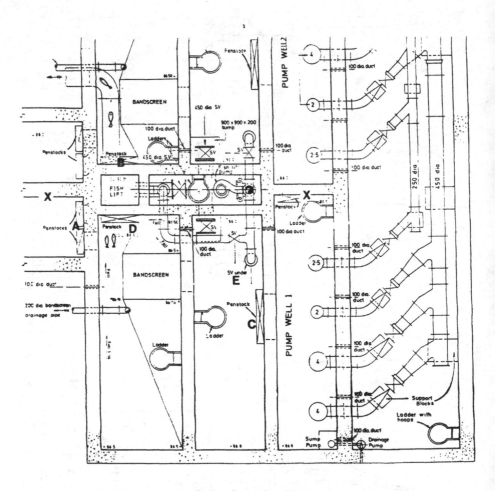

Fig 6   Downstream
Migration by Fish Lift

6.   CONCLUSIONS

The River Fane Flow Regulation Scheme comes on stream at a time when developments in Telemetry and transfer of information make a high degree of control possible to the overall advantage of the river environment. Indeed the design concept of the Scheme makes such control essential if abstractions from Lough Muckno are to be optimised so as to protect the amenity of the lake itself in the first instance and of the River Fane System as a whole. We have been particularly fortunate in that a 16 year record includes two severe droughts in 1975 and 1976 and two particularly wet summers in 1985 and 1986 so that the data bank is available to test our proposed Operations Policy against documented extreme historical conditions. A considerable amount of work on recession analysis has already been carried out, projections have been made of how the recession rate would change in the lower catchment following substantial regulation of Lough Muckno as demand grows. The particular needs of migratory fish to move upstream and downstream, sometimes simultaneously depending on species, during both the gravitational flow and pumping periods has resulted in a novel approach to Pumping Station Design and we are pleased to report that first recorded fish movements have taken place in October 1989.

Acknowledgement

The Author would like to acknowledge the permission of the Dundalk Urban District Council to present this Paper. He would also like to acknowledge the contribution of Associate Consulting Engineers, Rofe Kennard and Lapworth, Surrey, UK to the Project, as well as the three Contractors involved: SIAC Construction (Western Div.) Ireland, SPP (Projects) Ltd. Reading, UK and Kvaerner Boving Limited, Rotherham, UK.

References

1.   "The Modelling of Post Works Reservoir Behaviour using PreWorks Recorded Levels and Outflows". M. F. Garrick, paper presented to a joint Meeting of Water Engineering Section of I.E.I. and Rep. of Ireland Section of IWES February 1986.

2.   Institute of Hydrology 1980 Low Flow Studies Report No. 1 IOH Wallingford, UK.

3.   "River Recession Analysis in the Operational Management of Lake Abstraction" S. Walker and D. Pearson Jour. IWES Vol. 39, No. 3 1985.

4.   "Information for Reservoir Control in the Northumbrian Water Authority" J. Edwards and P. Johnson Jour. IWES. Vol. 32, No. 3 1978.

PART VII

Management of estuaries and beaches

Chapter 24

RIBBLE ESTUARY WATER QUALITY IMPROVEMENTS: AN INTEGRATED WASTE WATER MANAGEMENT PLAN TO ACHIEVE BATHING WATER COMPLIANCE

P C Head and D H Crawshaw (North West Water Ltd, Warrington, UK), S K Rasaratnam and J R Klunder (North West Water Ltd, Preston, UK)

As part of the United Kingdom's programme to bring its identified bathing waters up to the standard required by the EC Bathing Water Directive, North West Water has developed an integrated waste water management plan to improve the water quality of the bathing beaches situated at the mouth of the Ribble Estuary. This plan is based on an extensive investigation to determine the most appropriate way of dealing with both base flows (flows during dry weather or moderate rain), and storm flows (overflows from combined sewerage systems during wet weather). Mathematical models were used to examine the hydraulic behaviour of the sewerage systems and the subsequent dispersion of effluent into the receiving waters.

To determine the discharge characteristics of the existing sewerage systems discharging to the estuary, extensive hydraulic models were set up and validated by means of flow measurements gathered from critical points of the sewer networks. These models were then used, in conjunction with a time series rainfall record for the area, to investigate the effect of the intensity and duration of storm events on the volumes of sewage to be discharged. Estimates of the base and storm flows in the systems, derived from the hydraulic models, were used to determine the inputs to the dispersion model.

Dispersion modelling of various possible waste water management schemes, involving different degrees of sewage treatment and long sea outfalls, individually and in combination, in conjunction with tracer releases demonstrated that the use of a long sea outfall for all or part of the base flow was unlikely to lead to consistent compliance at the bathing waters situated at the mouth of the estuary. A suitable degree of compliance could be achieved by means of a sewage treatment scheme designed to remove at least 99 per cent of the bacterial load from the largest input.

INTRODUCTION

The well known seaside resorts of Southport and Lytham St Anne's, which are situated near the mouth of the Ribble estuary (see Figure 1) were not included in the original list of 27 bathing waters submitted by the UK Government to the European Commission in 1979 for designation under the Bathing Water Directive (Council of European Communities, 1976). This caused considerable local concern which led, in 1986, to a formal complaint being made to the Commission and a national review of how the Directive should be applied. As a result of this review, which identified more than 350 additional beaches, an extensive sampling programme was instigated to determine the extent to which these bathing waters complied with the microbiological standards contained in the Directive. This confirmed earlier studies which indicated that the waters off both Southport and Lytham did not consistently comply with the microbiological standards. North West

Water Treatment – Proceedings of the 1st International Conference, pp. 245–254

Water was asked by the government to determine how the quality of these bathing waters could be improved to ensure compliance.

Preliminary investigations of the location of sewage discharges to the area, and a detailed examination of ways to achieve compliance for the bathing waters off the nearby Fylde coast (Crawshaw & Head, 1989; Head et al., in press), showed that solutions to the problems at Southport and Lytham would only be possible within the context of an integrated programme to reduce polluting inputs to the Ribble estuary as a whole. It was apparent that, in addition to the effects of discharges of sewage and storm water near the bathing waters, the quality of the outer estuary could be influenced by discharges from Preston via the sewage works at Clifton Marsh, and possibly combined sewage overflows within the town itself.

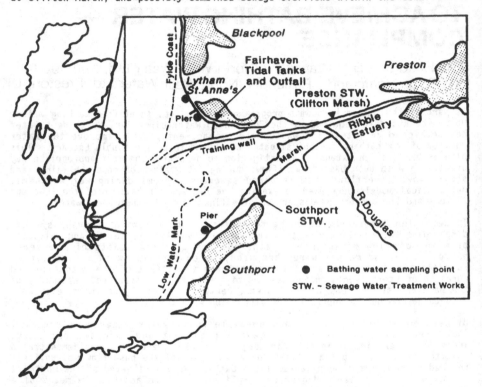

Figure 1.   Ribble Estuary showing bathing waters and existing sources of pollution

Additionally, sewage effluents from areas draining to the River Douglas were also liable to have an effect (Figure 1). It was important to ensure that any scheme to improve the condition of waters near the bathing areas did not result in a worsening of microbiological or other conditions elsewhere. In particular it was essential to ensure that the important salmon fishery in the estuary and further upstream was not adversely affected.   Investigations to determine the most appropriate scheme involved mathematical modelling and bacteriological investigations of the sewerage systems and receiving waters, along with tracer experiments to determine the significance of the existing inputs.

EXISTING CONDITIONS

The effects of the discharges from the most significant of the existing discharges to the estuary were investigated by means of a series of tracer releases carried out during 1988 and 1989.   In these, quantities of dye (rhodamine) and bacterial spores (Bacillus globigii) were added to the effluent from Southport Sewage Treatment Works (STW), Preston STW and the tidal sewage storage tanks at Fairhaven. In all, tracer tests were carried on 6 occasions in 1988 and 1989, involving 2 releases from Southport and Preston and 4 from Fairhaven. The main findings of these tracer releases were:

- discharges from Preston STW could be detected in the outer estuary in the waters off Lytham, but not off Southport;

- discharges from Southport STW were well dispersed in the outer estuary and were not detectable in the waters off either Southport or Lytham;

- discharges from Fairhaven Tanks were detectable in the waters off Lytham but not Southport, and during a period of strong south westerly winds could also be detected in water further north along the Fylde Coast.

From the determinations of the numbers of *B. globigii* found in the samples taken during the course of the tracer releases, it is probable that bacteria contained in the discharges from Preston STW and Fairhaven Tanks would lead to concentrations in the waters off Lytham in excess of those allowable by the Directive. The numbers of spores recovered from the waters off Southport and the Fylde Coast indicate that any effect from the above sources is small and unlikely to result in failure to comply with the Directive.

MODELLING OF THE SEWERAGE SYSTEM

Background and Verification
The hydraulic performance of each of the sewerage systems draining to the existing outfalls was examined in conjunction with the Water Research Centre (WRc) using the Wallingford Storm Sewer Package (WASSP) (DOE/NWC, 1983). With this package it is possible to simulate the effects of surcharging and flooding and to incorporate a wide range sewer ancillaries, including overflows and storage tanks. Information on the pipe networks and contributing areas was obtained from the local authorities. Surveys of flow and impermeable area were coordinated by WRc, and soil moisture deficit values for calculating the urban catchment wetness index were supplied by the Meteorological Office. The validity of the flows predicted by the model was checked by means of an extensive series of flow measurements carried out at key points on the sewer network. In all, data from some 50 flow monitors and 20 rain gauges were used in the validation process. The hydrographs produced by the WASSP modelling were used to calculate the inputs to the dispersion model for the various combinations of treatments investigated.

Storm Water Management
Since it is necessary to ensure that any scheme to achieve bacteriological compliance of the bathing waters during the bathing season must be capable of dealing with all but the severest storms (this is taken by the UK regulatory authority to include storm events of up to a 1 in 5 year return period) a considerable effort was directed towards modelling the effects of storms on the sewerage systems. This work provided estimates of the volumes of storm water liable to overflow from the existing sewerage systems under a wide range of rainfall conditions. Following on from the work previously carried out for the Fylde Coast studies (Crawshaw & Head, 1989; Head *et al.*, in press) hydrographs for use with the storm discharge dispersion model were obtained by running the WASSP models for Fairhaven and Preston with a sample of 44 summer storms from a time series rainfall suite representing the storms of a typical year, and 177 summer storms derived from rainfall data at Blackpool Airport. In the case of Southport, the most severe summer storms from the annual time series were used in conjunction with 25 summer storms derived from a local 15 year rainfall record.

DISPERSION MODELLING

Background
The model used to simulate the dispersion of effluent from outfalls to the estuary and coastal waters was developed by WRc. The dispersion processes are represented by a random walk procedure superimposed on the water movements derived from a hydrodynamic model covering the majority of the Irish Sea. The hydrodynamic model was validated by comparison with surface current information derived from Admiralty charts, Ocean Surface Current Radar deployments, and float track data (Crawshaw & Head, 1989).

In assessing the effects of the base flow, that is flows of up to about 6 x the dry weather flow (DWF), the dispersion model was run in a quasi steady-state manner with a run in period of up to 6 tidal cycles, to allow conditions to stabilize. An alternative form of the dispersion model was used to investigate the impact of storm water discharges. In this case there is no run in period and the effects of a discrete input of storm water is followed over a number of tidal cycles. The concentration of bacteria used for the inputs in the modelling exercises were estimated, wherever possible, from measurements of the existing discharges. Where data were not available, usually for storm water inputs, concentrations were taken to be half the estimate for the associated continuous discharge (Crawshaw, 1988). From the work carried out in relation to the Fylde Coast investigations it was known that die-off rates (t_{90}) for bacteria in the relatively turbid waters typical of the area could be in excess of 30 hours. For all the Ribble investigations the conservative value of 40 hours was used. With such a long die-off time the bacterial distributions produced by the model differ little from those resulting from dispersion alone.

The Existing Situation

From the regular sampling of the bathing waters from 1986 to 1989 it was apparent that the bacteriological quality was better at Southport (compliance for about 80 per cent of samples collected) than at Lytham (compliance for about 70 per cent of samples). In order to try to determine which of the various discharges to the estuary might be causing the poor quality of the bathing waters, the dispersion model was used to simulate the effects of the discharges singularly and in combination.   As can be seen from Figure 2, the effects of the sewage discharges to the Ribble estuary from Preston STW and Fairhaven Tanks lead to concentrations of bacteria in the vicinity of Lytham St Anne's sampling point which are in excess of those allowed by the Directive. Discharges from Southport STW have little effect on bacterial concentrations in the bathing waters off either Lytham or Southport, but the existing storm water discharges have a significant effect on the bathing waters at Southport (Figure 3).

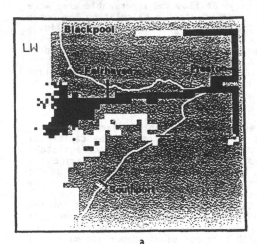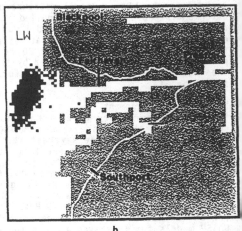

a   b

Figure 2.   Effects of the present discharges from Preston and Fairhaven on bacterial concentrations.   Dispersion model simulations, ▬▬▬ E. coli concentrations greater than 2 000 per 100 ml at low water: a Preston STW, b Fairhaven Tidal Tanks

Options for Treating the Base Flow
North bank

From the investigations into the existing situation, it was apparent that to achieve the required bacteriological quality at Lytham in dry weather would require improvements to the present sewage disposal arrangements for both Fairhaven and Preston. Measurements of bacterial concentrations in the inflowing rivers showed them to be appreciable. Model runs of the river inputs indicated a significant effect on bacterial distributions in the upper and middle estuary but that their effects did not extend into the bathing waters. In view of this the

investigations into the most appropriate method of reducing bacterial loads to
the estuary, to achieve bathing water compliance, concentrated on dealing with
the contributions from Preston and Fairhaven. Because of the various constraints
imposed by the existing land use around the estuary the available options could
be narrowed down to:

- a sea outfall for flows from Fairhaven or Fairhaven and Preston
  together;

- treatment of Fairhaven and Preston flows separately and discharge via
  the existing outfalls;

- transfer flows from Fairhaven to Preston for combined treatment with a
  discharge via the existing Preston outfall.

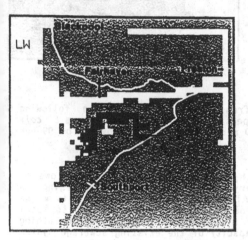 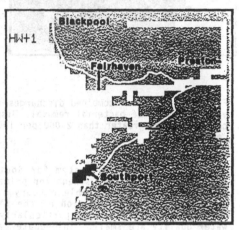

a   b

Figure 3. Effects of the present storm discharges from Southport on bacterial
concentrations. Dispersion model simulations, ████████ E. coli concentrations
greater than 2 000 per 100 ml: a Southport STW, b Southport Combined Sewer
Overflows.

In considering possible locations for long sea outfalls there was only one
possible site, situated to the north of Lytham, where it would be feasible to
site a headworks within a reasonable distance of water sufficiently deep to
provide adequate initial dilution of the effluent. Even from this position an
outfall of around 6.5 km would be required to reach water with a low water depth
greater than 10 m.

The probable effects of variations of these schemes were compared with the
assumption that 3DWF would be discharged. Investigations of the effects of
dealing with flows of around 6DWF were left until the modelling of the effects of
storm discharges with 3DWF receiving treatment had been examined. Removal of
the Preston and Fairhaven flows from the estuary to a long sea outfall resulted
in a dramatic improvement in the bacteriological quality in the bathing waters
off Lytham, but produced an area of high bacteriological contamination off the
Fylde Coast near to a number of identified bathing areas. Even the discharge of
flows from Fairhaven alone resulted in a relatively large area of high bacterial
numbers near to these sensitive high amenity waters.

The effects of treating the Fairhaven and Preston flows, either separately, or
together at the present Preston STW site, depend on the assumptions made about
the degree of bacterial removal. The bacterial load from Preston is so large
in relation the volume of the receiving waters that it is necessary to reduce
bacterial concentrations by 99 per cent before the effects on the Lytham bathing
waters become acceptable (Figure 4). If Fairhaven flows are treated separately
and discharged on the ebb a bacterial reduction of between 90 and 95 per cent
would be required.

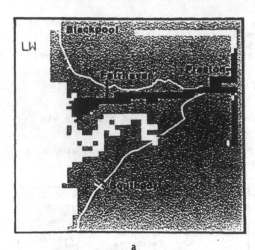 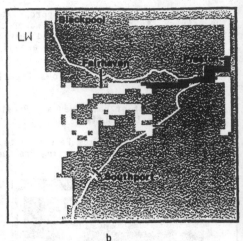

a   b

Figure 4. Effects combined discharges from Preston and Fairhaven following various levels of bacterial removal. Dispersion model simulations, ▬▬▬ *E. coli* concentrations greater than 2 000 per 100 ml: a   95 per cent removal,   b   99 per cent removal.

South bank
The existing drainage system for Southport is designed to convey flows of combined foul and storm sewage for primary treatment at a works discharging via Crossens Pool to the Ribble estuary. Although the effluent from this works comprises a large proportion of the flow in Crossens Pool, leading to local oxygen depletion problems, particularly in the summer, its effects on bathing water quality are small.   Inadequate capacity of the existing sewerage system leads to premature operation of existing combined sewer overflows and the flooding of properties.   In planning improvements to the system to alleviate these problems, it was necessary to determine how best to deal with the base flow so that the effects of discharges of storm water on bathing water quality could be reduced to a level compatible with the requirements of the Bathing Water Directive.   Since only part of the system involves combined overflows discharging to areas near the bathing waters, it was necessary to consider conveying different multiples of the base flow to the site of the existing works, depending on   whether the overflows were to inland water courses or to the sea.   Additionally the possibility of constructing a long sea outfall for some or all of the base or storm flows needed to be considered.   After taking into account the various constraints associated with existing land use and obtaining permissions for new constructions it was possible to narrow the options to be considered to:

- convey all flows up to a 1 in 5 year storm, from areas where overflows would need to be to the sea, to the existing sewage works site.   Allow volumes in excess of these to be discharged via a relatively short outfall (about 1 km) near the bathing waters. For areas with overflows to inland water courses, transfer approximately 6DWF to the existing sewage works site.   Discharge flows in excess of 6DWF arriving at the works via an overflow to Crossens Pool;

- convey all flows from areas where overflows would need to be to the sea to a coastal site for pretreatment and possibly storage before discharging to the sea.   Discharge volumes up to a 1 in 5 year storm via a long outfall (about 4.5 km) and those in excess of this via a short outfall (about 1 km) near the bathing waters.   Transfer flows of up to 6DWF from inland areas to the existing sewage works site;

- convey flows up to 6DWF from all areas to the existing sewage works site.   Discharge flows of between 6DWF and a 1 in 5 year storm via a long sea outfall (about 4.5 km), with or without storage to limit the time of the discharge; or store and transfer to the works at a rate of 6DWF. Discharge flows in excess of those for the 1 in 5 year storm to sea via a short outfall (about 1 km) near the bathing waters.

For all these options it was necessary to determine the amount of storage that would be required to limit the number of discharges from any combined sewer overflows, to inland or marine waters, to a frequency that would be acceptable in terms of general environmental impact, and in the case of those to sea to that which would not prejudice bathing water quality.

Options for Treating Storm Flows
North bank
For Fairhaven, investigations concentrated on determining which storms would lead to spills from the tanks outside the 3 hour period following high water (the tidal window), when discharges receive maximum dilution and tend to move offshore away from the bathing area. This involved modelling the effect, in terms of probable spill volumes, of all the time series summer storms, starting at hourly intervals through the tidal cycle. After combining some of the output to take account of similar spill volumes from different storms starting at different tidal states, an analysis of rainfall data from nearby Blackpool airport (Head et al., in press) was used to determine the number of spills to be expected during an average summer. From Figure 5 it can be seen that increasing the existing storage from 10 000 m³ to around 25 000 m³ would dramatically reduce the number of spills liable to occur. Above this figure the benefit in terms of a reduction in the numbers of spills diminishes. Increasing the tidal window from 3 to 4 hours has a relatively small effect on spill frequency, whereas the removal of 3DWF for treatment elsewhere would greatly reduce the number of spills.

A much more limited examination of the storm water management with regard to receiving water quality was required in the case of Preston, as the initial investigations showed that the bacterial loads from the combined sewer overflows on the sewerage system only affected the upper and middle estuary, and that significant numbers of bacteria did not reach the bathing areas.

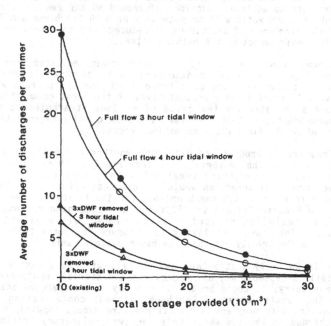

Figure 5   Relationships between storm water storage requirements and number of discharges during an average summer for Fairhaven Tanks

South bank
Because the only locations for overflows of storm water from the coastal areas of Southport are near to the bathing waters, the aim of any storm water management

scheme is to contain all flows up to a 1 in 5 year storm, and discharge flows in excess of this via a short outfall that will not result in gross pollution of the bathing beaches. Additionally it is necessary to ensure that flows of between 6DWF and the 1 in 5 year storm, which would exceed the volumetric capacity of the sewage works, are discharged at a location and a frequency which will not prejudice other uses of the estuary.

Investigations are proceeding to determine the storage capacity necessary to contain a 1 in 5 year storm, and the effects of discharging the volumes stored to the sea via a long sea outfall or the existing sewage works. In determining the best overall solution for the discharge of the storm water, it will be necessary to ensure that transfer to the sewage works at rates greater than 6DWF will not result in an excessive number of discharges into Crossens Pool.

Although the investigations are not yet complete, sufficient information is available to ensure that the options are properly evaluated in terms of storage and pumping capacity required, and probable effect on the receiving waters.

By ensuring that sufficient storage is provided to contain a 1 in 5 year storm, any samples reflecting contamination of the bathing waters resulting from discharges via the short outfall can be excluded from assessing compliance, on the grounds of abnormal weather conditions.

DISCUSSION

The investigations so far carried out into the most appropriate way of improving the quality of the bathing waters to that required by the Bathing Water Directive have shown that improvements will be required for the disposal of both the base and storm water flows.

In the case of the base flow, the main improvement required is to achieve a reduction in the bacterial input from Preston STW of the order of 99 per cent, and to ensure that the discharge from Fairhaven is either incorporated in this or treated separately to achieve reductions of around 90 per cent. Improvements to the treatment of storm water will be necessary at both Fairhaven and Southport to ensure that the frequency of discharges is reduced, and that they are located so as to minimize their impact on the bathing waters.

The final scheme to achieve the degree of improvement indicated from the studies can only be determined after discussions with the regulatory authority responsible for authorizing new or altered discharges to surface waters, (the National Rivers Authority) and representatives of the local communities. Studies are in progress to determine the costs of various strategies to achieve the required improvements, so that the overall economic and environmental costs can be incorporated in the final decision making process.

Investigations are in progress to determine whether a consistent reduction of around 99 per cent in the bacterial load from Preston STW could be achieved by means of conventional secondary treatment with a long retention period, or whether some form of disinfection would be more cost effective. These studies, which are also relevant to treatment options for the Fairhaven flow if the costs of transfer to Preston STW are significant, are examining various methods of disinfection, including peracetic acid, ozone, ultra-violet light and flocculation. These are less likely to produce environmentally undesirable by-products than the traditional chlorine based disinfectants.

Although it is unlikely that a long sea outfall will be a cost effective and environmentally acceptable way of reducing the bacterial inputs from the north bank of the estuary, it could be a suitable way of discharging screened storm water from the coastal areas of Southport without contaminating the bathing waters. However, with many environmental pressure groups opposed, on principle, to any discharges to the sea which only receive preliminary treatment, even a proposal for an intermittent discharge such as this could cause controversy.

In designing any sewage treatment scheme, it is necessary to determine the combination of treatments for base flows and storm water which results in the least overall impact on the environment at an economic cost. By combining hydraulic modelling of sewerage systems, by means of packages such as WASSP, with hydrodynamic and dispersion models of the receiving water, it is possible to examine the range of options available in far greater detail than was the case previously. Provided both types of model are adequately validated against field data, the task of achieving the desired degree of compliance with receiving water standards becomes much more one of objective scientific investigation, rather than one based to a very large extent on intuition and experience.

CONCLUSIONS

When putting forward proposals to improve water quality it is important that the cause of the problem is sufficiently well understood, to enable the consequences of various possible improvement schemes to be properly evaluated.

Mathematical models of the sewerage system and the receiving waters, used in conjunction with adequate investigative work, provide a powerful tool for illustrating the probable consequences of various courses of action. Once it can be established that a number of different schemes will result in the degree of improvement initially required and, in most cases, what would be required to achieve a greater improvement, the decision as to which is appropriate to any particular location will then involve weighing the costs of each option against the immediate and future environmental benefits.

ACKNOWLEDGEMENT

These investigations involved contributions from a number of people working for NWW and its consultants. The authors would like to thank them for their help in formulating the problems to be addressed and evaluating possible solutions.

REFERENCES

Council of European Communities. 1976. Council Directive of 8th December 1975 concerning the quality of bathing water. (76/160/EEC) *Official Journal of the European Communities* **No. L31**, 1-7.

Crawshaw, D H. 1988. Mathematical modelling of sewage dispersal and scientific studies. Appendix H. Fylde Coastal Bathing Water Improvements. Planning Study Final Report. North West Water, Warrington.

Crawshaw, D H and Head, P C. 1989. Fylde Coast bathing water improvements - environmental investigations for the design of sea outfalls. In: *Long Sea Outfalls*. Thomas Telford, London, 75-88.

Department of the Environment/National Water Council. 1983. The Wallingford Procedure. *DoE/NWC Standing Technical Committee Report* **No. 28**

Head, P C, Crawshaw, D H and Rasaratnam, S K. In press. Fylde Coast bathing water improvements - storm water management for compliance with the Bathing Water Directive. Proceedings of 2nd Wageningen Conference - Urban Storm Water Quality and Ecological Effects upon Receiving Waters. 20-22 September 1989. Pergamon Press, Oxford.

Chapter 25

ADVANCES IN WATER TREATMENT AND ENVIRONMENTAL MANAGEMENT: A STUDY OF POLLUTION CONTROL AT A STORMWATER OVERFLOW

N J Bennett and R V Farraday (Balfour Maunsell Ltd, UK)

In the United Kingdom pollution of beaches and inland waterways has become a serious environmental issue demanding urgent attention. Pressure from the general public and from EC legislation is mounting on authorities to improve the present level of service.

Stormwater overflows are regarded as transient, intermittent point sources of pollution. There are numerous reports which conclude that pollution arising from stormwater overflows can account for the majority of pollution problems experienced in water courses and on beaches.

Southern Water plc are addressing the problem of the polluting influence of stormwater overflows on receiving waters by installing Storm King Overflows for removing solids from sewage and stormwater. A Storm King Overflow has no moving parts, needs no power to operate and is self-cleansing. Southern Water has installed five such devices on a sewerage system in Bexhill, Sussex England to control and screen stormwater before discharge into an adjacent stream. Balfour Maunsell Ltd. has been commissioned to evaluate the "treatment" efficiency of one of these installations.

The research programme involves monitoring the frequency of discharge and determining the long term effects of discharge on the receiving stream. Efficiency is being measured both in chemical terms, using automatic sampling devices on the inlet and outlet pipes, and in aesthetic terms by collecting from the stream, material which can be described as being of sewage origin. Flows through the Storm King Overflow and in the stream are also being recorded.

The long term effect of storm discharges on the stream is being measured in biological, bacteriological and chemical terms. Biological analysis examines the

© 1991 Elsevier Science Publishers Ltd, England
Water Treatment – Proceedings of the 1st International Conference, pp. 255–270

presence of invertebrate fauna using the BMWP score and ASPT techniques. Bacteriological and chemical sampling involves the measurement of coliforms, faecal streptococci, BOD, suspended solids and ammonia from five sites along the receiving stream.

The project is ongoing and the findings to date are reported in this paper.

1.   INTRODUCTION

In the United Kingdom pollution of beaches and inland waterways has become a serious environmental issue demanding urgent attention. The demands of EC directives have increased public awareness of bacteriological and other standards to be achieved at bathing beaches and there has been a lot of activity on the control of pollution in the environment. Stormwater overflows are regarded as transient, intermittent point sources of pollution. There are numerous reports [1] which conclude that pollution arising from stormwater overflows can account for the majority of pollution problems experienced in watercourses and on beaches. But, more information is required on the effects of stormwater on the environment.

2.   PRESENT PRACTICE

In general, standard practice in the United Kingdom concentrates on the function of a stormwater overflow as a "flow splitter" to remove excess volumes of storm sewage from a foul sewer with inadequate hydraulic capacity. In consequence, there has been much attention paid to stormwater overflow settings.

For example, the recommendation by the original Ministry of Health of setting storm overflows at 6 x DWF, providing full treatment for 3 x DWF and settlement for flows from 3-6 x DWF in storm tanks regardless of the somewhat obscure origins of these recommendations, have been in use for many years.

This recommendation was modified on slightly more rational grounds by the Technical Committee on Storm Overflows [2], with the production of Formula 'A'- commonly regarded as the present standard.

The development of computers and advanced mathematical techniques has enabled further refinement and it is now normal to assess a sewerage system and the impact of an overflow on the hydraulics of a system using a design package such as WASSP or WALLRUS.

The question of the polluting influence of stormwater overflows on the receiving waters environment has been basically neglected.

It is obvious that, in general, the mass pollutant escape from a stormwater

overflow is the most damaging to the environment and is a function of the volume
of water overflowing to the receiving waters.

Some attention has been paid to the avoidance of the discharge of large and
aesthetically offensive materials by the use of screens on overflows, and it is
common to use 12 - 25mm spacing bar screens to retain larger solids in the flow
passed for treatment. It is also usual to incorporate some form of dip plate to
retain floating solids within the overflow and to discharge them back to the
sewer at the end of the storm when water levels recede.

3.   PRESENT TRENDS

It is clear that there will be a revolutionary change in the way stormwater
overflows are considered in the future. Present research is revealing the
deficiencies of existing practice and it should produce the techniques which will
be needed to rectify these deficiencies. Pressure from the general public and
from E.C. legislation is mounting on Authorities to improve the present level of
service. This trend is reflected in other countries too.

4.   SOUTHERN WATER

Southern Water plc are addressing the problem of the polluting influence of
stormwater overflows on receiving waters by installing Storm King Overflow
(SKO) devices for removing solids from sewage and stormwater.

Three sets of SKO's have been constructed by Southern Water on a combined
sewerage system in Bexhill to control, attenuate and screen the discharge of
stormwater from the drainage area. Following the commissioning of these devices
in September 1987, Southern Water instigated a research project to study the long
term effects of these stormwater discharges on the receiving water - the Egerton
Park Stream. This study was carried out in-house by Southern Water until
November 1989, when the newly formed company, Balfour Maunsell Ltd., was
appointed as Consultant on the project.

5.   THE STORM KING OVERFLOW

This device is a hollow cylindrical vessel with a dished base and a flat top, see
Fig 1. The geometry of the interior and the shape and placing of the entry jet
(A) are carefully arranged so that flow entering the device must follow a
predetermined path through the vessel.

Sewage is tangentially fed into the side of the cylinder (B) so that the contents
rotate gently about the vertical axis. The flow first spirals gradually down the
perimeter allowing time for solids to settle out by gravity, aided by the drag
forces of the boundary layer at the wall and bottom surfaces of the cylinder.

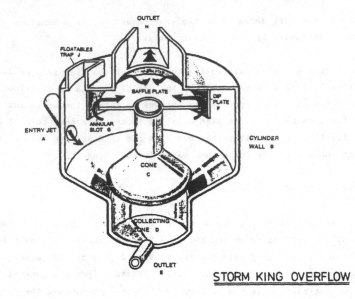

FIG 1   STORM KING OVERFLOW

In the bottom of the vessel, a cone (C) is mounted so that its edge is clear of the base; the collecting zone (D) for the separated solids is located under the cone discharging solids to outlet (E).

Having reached the bottom, the main flow is then directed by the shape of the vessel away from the perimeter and up the middle as a broad spiraling column; but, it now rotates at a slower velocity than the outer downward flow. The interface between the outside downward circulation and the internal upward circulation is called the shear zone, where there is a sudden and marked difference of velocities. This differential creates a coagulating and flocculating effect, encouraging further separation of the solids. The location of the shear zone is critical and it is precisely positioned by a dip plate (F). When the flow reaches the top of the device, the decanted liquid is discharged through an annular slot (G) to the outlet (H). Floating material is captured in the floatables trap (J) at the top of the device, between the wall of the cylinder and the dip plate.

6.   THE EGERTON PARK STREAM

The Egerton Park Stream is a small urban stream which rises to the north of Bexhill, flows through the centre of the town and then discharges onto the beach to the south of the town, see Fig. 2. The stream was culverted and widened during the installation of the SKO's, to cater for the extra flow from the catchment.

Several natural springs, of which many are rich in iron, seep into the stream along its length. In its upper reaches the stream also flows alongside a recreation ground which used to be a refuse tip.

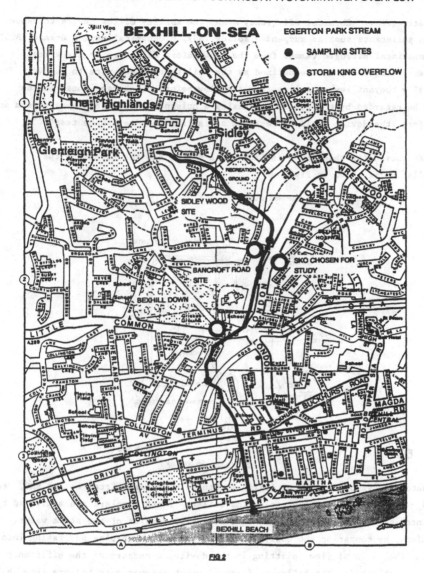

FIG 2

The consent standard imposed on the discharge from the SKO's was that"the
effluent shall consist only of storm sewage".

7.   THE STUDY

The study was designed to

a)   evaluate the "treatment" efficiency of a SKO.

b)   measure the long term effects of stormwater discharges on the stream itself.

7.1 Efficiency of a Storm King Overflow
The efficiency of a SKO can be looked at in two ways - in chemical terms and

aesthetic terms. Chemical efficiency means the ability of the device to remove
common pollutants such as suspended solids (SS), biochemical oxygen demand (BOD)
and ammoniacal nitrogen (AmmN) from the flow before discharge to the stream.
Aesthetic efficiency can be defined as the ability of the device to remove
neutrally buoyant sewage solids from the flow before discharge; these solids
could be regarded by the public as causing a nuisance in the receiving stream and
therefore, the device could be regarded as failing the "telephone test" [3].

The SKO chosen for this study is located in Fig 2. and the layout of the
installation is shown in Fig 3.

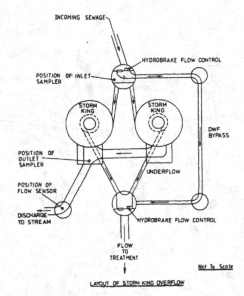

FIG.3

LAYOUT OF STORM KING OVERFLOW

Two automatic sampling devices (Dalog machines) have been fitted to the SKO, see
Fig 3 - one on the inlet and one on the outlet. These samplers are activated by
the incoming and outgoing flow and take composite samples of the storm water.
Therefore, by comparing the quality of inlet and outlet samples and taking into
account the ratio of flow splitting in the device, a measure of the efficiency
can be established. The following formula, which assumes mass balance through
the the device, applies.

Efficiency $E = \dfrac{N/R}{(1+N/R)} \times 100\%$

where N/R = $\dfrac{\text{Mass of Pollutant in Underflow}}{\text{Mass of Pollutant in Overflow}}$

N = $\dfrac{\text{Pollutant Concentration in Underflow}}{\text{Pollutant Concentration in Overflow}}$

and R = $\dfrac{\text{Rate of Overflow}}{\text{Rate of Underflow}}$

Simultaneous with recovering the above samples and after the storm has
passed, an aesthetic survey is carried out upstream and downstream of

the SKO discharge point. In this survey, all material which can be described as being of sewage origin is collected in bags, counted, described, recorded and then disposed of.

7.2 Flow Measurement

Automatic flow recorders have been installed in the SKO and in the Egerton Park Stream. Data are downloaded from these recorders at regular, fortnightly, intervals and reproduced in graphic and tabular form for further analysis. Rainfall is recorded at Bexhill and data here are retrieved at similar intervals.

From these measurements, inter-relationships can be established and an overall assessment of the conditions under which the SKO's operate can be defined.

7.3 Measurement of the Long Term Effect on the Stream

This part of the study involves three sampling routines to measure bacteriological, chemical and biological levels in the stream.

Water samples are taken from the stream each month and analysed for bacteria and chemical levels. Bacteriological and chemical analysis involves the measurement of Total Coliforms, E. Coli, Faecal Streptococci, BOD, SS, AmmN and pH. Bacteriological and chemical analysis enables short term polluting effects to be assessed.

Every three months, biological sampling is undertaken in the stream bed to examine the presence of invertebrate fauna. Biological analysis measures the existence of invertebrate fauna according to the Biological Monitoring Working Party (BMWP)[4] score and Average Score Per Taxon (ASPT) [5] techniques. A method called the Lincoln Quality Index (LQI)[6], which combines both the BMWP score and ASPT, is then used in such a way as to give an overall quality rating or index for the stream. Invertebrate sampling is a good method of demonstrating organic pollution or stream life decay over long periods.

Five points were chosen along the stream from which to take samples - two upstream of the SKO, to act as control points, and three downstream, see Fig 2. Also, during periods of storm, when the SKO is operating, extra sampling (except biological) is carried out.

8.   THE RESULTS

8.1 General

No biological, bacteriological or chemical quality data were available for the stream prior to the commencement of this study. Therefore, no

direct comparison of previous water quality can be made with the
results of this study. Also, automatic flow sensors in the stream and
in the SKO were not installed until the autumn of 1988.

Due to the quantity of information available from this study, this
paper has only presented results from two sampling sites - one upstream
(SIDLEY WOOD) and one downstream (BANCROFT ROAD) of the SKO under
scrutiny.

8.2 Biology

Under the LQI system [6], all sampling sites must be classified
according to their "habitat" before any interpretation of the BMWP and
ASPT scores is carried out. The sampling points along the Egerton Park
Stream are classified as "habitat-poor riffles" and, as such, enhanced
ratings are assigned to this category since this type of environment
will support a poorer invertebrate community even in the absence of
pollution.

The results from the two upstream or control sites suggest that the
quality rating is consistent, whereas the quality rating of the
downstream sites seems to fluctuate and generally improve when there is
no rainfall or storm spill events. Fig 4. shows the results from an
upstream site at Sidley Wood and a down stream site at Bancroft Road.

The diversity of flora is greater downstream of the overflow probably
due to the effect of the discharges. The invertebrate data show the
stream population to be unstable and, overall, according to the Lincoln
Quality Index, the stream itself has maintained a poor quality since
the start of the study and has not deteriorated as a direct result of
the spill events. Further longer term effects will be available as the
study progresses.

8.3 Chemistry

Urban run-off and the discharge from the SKO seem to be the main
causative factors in the variation of chemical quality of the stream.
Considering two sampling sites, one upstream (Sidley Wood) and one
downstream (Bancroft Road) of the SKO, the average value of the BOD for
the study (Dec 87-May 89) for each site was 2.57 and 3.89 mg/l
respectively - see Figs 5 and 6.

The average value of the SS content for each site was 22 and 52mg/l
respectively - see Figs 5 and 6.

Ammonia is toxic to fish and if a stream has ammonia levels of over
1mg/l it is considered not suitable for fish life. The average value

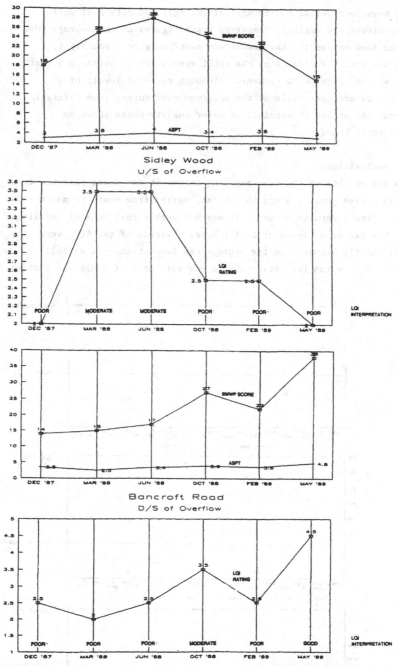

of ammoniacal nitrogen for the study for each site was 0.19 and 0.23 mg/l respectively. This confirms the fact that local residents have reported seeing eels swimming around the overflow position. However, when the overflow operates, BOD, SS and AmmN levels in the discharge

have been recorded at 4.9-106mg/l,52-281 mg/l and 2.19-8.92 mg/l
respectively, see Table 1. Comparing these figures to the average BOD,
SS and AmmN values for the downstream sampling site , see Fig 6,
suggests that the effects of the spill events are transient and that
the stream is quick to recover. Although elevated levels of
pollutants are detectable at the upstream site during some rainfall
events, the effect of rainfall on water quality there shows no
detrimental trend.

8.4 Bacteriology

From the results at the five sampling points it was found that bacteria
counts varied greatly along the stream, varied from month to month and
varied from season to season. It was noticeable that on most sampling
days the bacteria levels at Sidley Wood, upstream of the SKO, were
significantly higher than the other sites (except the sea outfall
site). One reason for this could be the existence of a disused rubbish

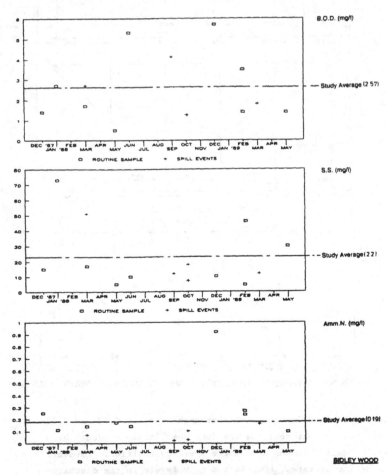

FIG 5   SUMMARY OF CHEMICAL ANALYSES FROM UPSTREAM SAMPLING POSITION

tip (now a recreation ground) situated alongside the stream. This site
is also a popular place for dogs.

When operating, the SKO discharges a large number of bacteria
(typically 3×10^{6} coliforms/100ml) to the stream. But, spill events
are intermittent and, from the bacteria levels recorded at the Bancroft
Road site, (typically 6×10^{5} coliforms/100ml) coincide with 5 or more
dilutions in the stream. Of interest, though, is the fact that the
Egerton Park Stream flows out onto a designated bathing beach, see Fig
2. So, in this respect, bacteria counts are important. But, despite
the fact that bacteria counts can be difficult to interpret and are
highly variable, Bexhill's beach has passed the EC Bathing Water
standards. Also, the bacteriological analysis supports the evidence
from the chemical analysis that the effects of the spill events are
transient and that the stream is quick to recover. Results are shown
in Figs 7 and 8.

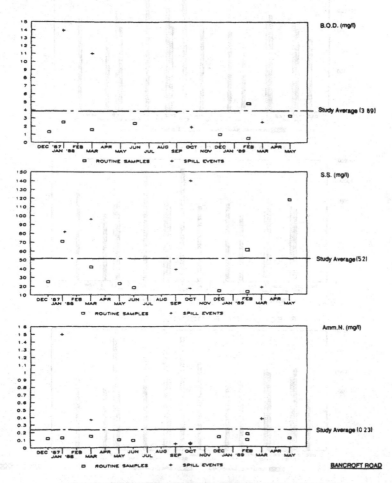

FIG 6   SUMMARY OF CHEMICAL ANALYSES FROM DOWNSTREAM SAMPLING POSITION

8.5 Rainfall

Rainfall has been measured since the commencement of the study and data
are presented in Fig. 9. The rainfall which occurred 3 days before a
routine sample and a spill event is shown. Total monthly rainfall is
not shown. Generally, it can be seen that elevated levels of
pollutants and bacteria occur during rainfall; but, that, these return
to normal when the storm has passed.

8.6 Overflow Efficiency

Six discharge events have been recorded during the study but, only on
one occasion (17/3/89), has inlet and outlet quality data been
retrieved simultaneously. From these data and using the formula
suggested in section 7.1, the treatment efficiency of the SKO has been

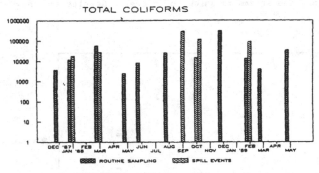

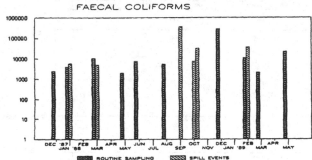

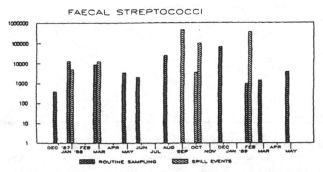

SIDLEY WOOD

FIG 7   SUMMARY OF BACTERIOLOGICAL ANALYSES FROM UPSTREAM SAMPLING POSITION

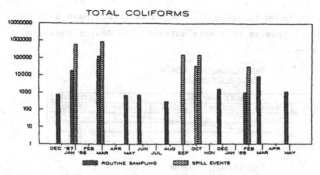

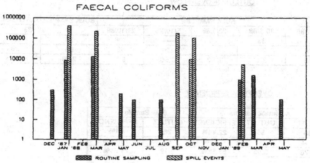

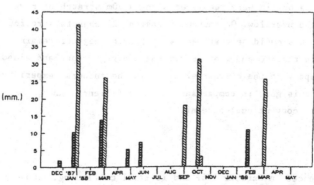

FIG 8 SUMMARY OF BACTERIOLOGICAL ANALYSES FROM DOWNSTREAM SAMPLING POSITION

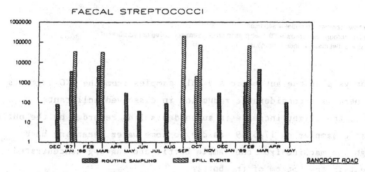

FIG 9   RAINFALL

calculated and is shown in Table 1. The calculations have been based on the fundamental premise that mass entering the device equals mass leaving the device.

SAMPLE ANALYSES FROM INLET OF OVERFLOW

DETERMINAND (mg/l)	BULK SAMPLE					AUTO SAMPLE
	29/1/88	30/3/88	23/9/88	5/10/88	21/10/88	17/3/89
BOD			NO DATA			65
SS						254
AmmN						9.76

SAMPLE ANALYSES FROM OUTLET OF OVERFLOW

DETERMINAND (mg/l)	BULK SAMPLE					BULK SAMPLE	AUTO SAMPLE
	29/1/88	30/3/88	23/9/88	5/10/88	21/10/88	17/3/89	
BOD	27	33	4.9	—	106	26	56
SS	72	73	65	186	52	83	281
AmmN	3.5	3.1	8.92	3.56	8.3	2.19	3.95

OVERFLOW EFFICIENCY

DETERMINAND (mg/l)	Automatic Sample			Bulk Sample	
	INLET	OUTLET	EFFICIENCY	OUTLET	EFFICIENCY
BOD	65	56	22.46%	26	64%
SS	254	281	0.43%	83	70.59%
AmmN	9.76	3.95	63.58%	2.19	79.81%

Average flow recorded on outlet = 540 l/sec
Average flow through underflow = 60 l/sec (based on performance curve of hydrobrake)
Inlet mass = outlet mass + underflow mass

TABLE 1

The analysis of the automatic and bulk samples from the SKO suggests that there is a considerable reduction in dissolved pollutants. However, the slight increase in suspended solids recorded by the outlet automatic sampler on 17/3/89 was due to some water decanting back through the machine from the sample bottle and leaving concentrated sediment in the bottom of the bottle.

Two aesthetic studies have been carried out. On one occasion 53 objects of sewage origin were recovered along a 50m stretch of river downstream of the overflow. On another occasion, 26 objects were found. One reason for this could be that neutrally buoyant objects remain suspended within the SKO and when conditions are right (perhaps aided by leaves) escape with the discharge. But, in the authors' experience, this performance is good in comparison to the efficiency and effectiveness of conventional screens.

9. CONCLUSIONS

The study commenced in October 1987 and is still continuing. More invertebrate data are required to assess the long term trends of water quality in the stream; but, as suggested by the chemical and bacteriological analyses, despite there being transient effects from the discharges, the stream quickly recovers.

Regarding stormwater quality, the results suggest that the SKO improves the chemical quality of the discharge but does little to reduce bacteria. Also, from the results of one spill event (17/3/89) the device demonstrates a high treatment efficiency despite the fact that some solids are discharged to the stream.

REFERENCES

1.   Clifforde I.T, Saul A.J, Tyson J.M. (1986) - Urban Pollution of Rivers - the UK Water Industry Research Programme. International Conference on Water Quality Modelling in the Inland Natural Environment, Bournemouth .

2.   Technical Committee on Storm Overflows and the disposal of storm sewage - Final Report (1970). Ministry of Housing and Local Government, HMSO.

3.   Ellis J.B. (1989). The management and control of urban runoff quality. J.IWEM, 3 (No2). Discussion by N.J. Bennett, pages 123-124.

4.   Chesters R.K. (1980) Biological Monitoring Working Party. The 1978 national testing exercise. Department of the Environment Water Data Unit, Technical Memorandum 19.

5.   Armitage P.D et al (1983). The performance of a new biological water quality score system based on macroinvertebrates over a wide range of unpolluted running - water sites. Wat. Res, 17, 333-47.

6.   Extence C.A et al. (1987). Biologically based water quality management. Environmental Pollution, 45, 221-236.

Printed in the United States
by Baker & Taylor Publisher Services